T0144138

Crowd-Assisted Networking and Computing

Crowd-Assisted Networking and Computing

Edited by
Al-Sakib Khan Pathan

CRC Press
Taylor & Francis Group
Boca Raton London New York

CRC Press is an imprint of the
Taylor & Francis Group, an **informa** business

AN AUERBACH BOOK

CRC Press
Taylor & Francis Group
6000 Broken Sound Parkway NW, Suite 300
Boca Raton, FL 33487-2742

© 2019 by Taylor & Francis Group, LLC
CRC Press is an imprint of Taylor & Francis Group, an Informa business

Library of Congress Cataloging-in-Publication Data

Names: Pathan, Al-Sakib Khan, editor.
Title: Crowd assisted networking and computing / editor, Al-Sakib Khan
Pathan, Ph.D.
Description: Boca Raton, FL : CRC Press/Taylor & Francis Group, 2019. |
Includes bibliographical references and index.
Identifiers: LCCN 2018017151| ISBN 9781138294769 (hb : alk. paper) | ISBN
9780429453731 (ebook)
Subjects: LCSH: Crowdsourcing.
Classification: LCC QA76.9.H84 C745 2019 | DDC 004.6/5--dc23
LC record available at https://lccn.loc.gov/2018017151

Visit the Taylor & Francis Web site at
http://www.taylorandfrancis.com

and the CRC Press Web site at
http://www.crcpress.com

Dedicated to my parents (father, *Abdus Salam Khan Pathan*, and mother, *Delowara Khanom*) and my family (wife, *Labiba*, and two daughters, *Rumaysa* and *Rufaida*)

– *Al-Sakib Khan Pathan*

Contents

Preface

The plethora of mobile devices in the hands of people, generally termed the "crowd," has allowed us to envision various platforms of communications taking the same networked environment into consideration. Crowd computing, crowdsourcing, crowd-associated network (CrAN), and crowd-assisted sensing are some of examples of the crowd-based concepts where we like to harness the power of the people out in the web or connected via a web-like infrastructure to do tasks that are often difficult for individual users or computers to do alone. For instance, the key concept of an infrastructure-less network architecture named CrAN is the utilization of the crowd to complete the communication gaps among the associates. In CrAN, two types of components are involved: (1) *dedicated agents* and (2) *nondedicated agents*. The dedicated agents are those agents that are solely installed in the network to perform some specific tasks. In general, these agents are static and exchange information with nondedicated agents to achieve the networking goal. On the other hand, the crowd is the latter type of agent, who is equipped with necessary devices and acts like an intermediate relay in the network architecture. The crowd completes the communication gaps among the dedicated agents and thereby enables them to function properly.

As the technological landscape is constantly changing, in the near future, many of our communications will heavily rely on the crowd, who could be socially connected via the Internet or an Internet-like infrastructure. Gone are those days when people were relying only on selected sources of information, like traditional newspapers or TV channels. Today, we see, the crowd could supply information from distant and difficult-to-reach places faster than formal news and information channels. As the communications technologies and application software technologies advance, a single individual could represent several devices that could generate or supply information. Given this scenario, many challenging issues come forward, like assessing the reliability and correctness of crowd-generated information, the delivery of data and information via the crowd, middleware for supporting crowdsourcing and crowd computing tasks, crowd-associated networking and its security, and quality of information issues. This book aims to compile the latest advancements in the relevant fields.

After receiving several proposals for chapters from potential contributors from around the globe, a total of 10 chapters were accepted for the book. In these chapters, 27 authors have covered various aspects of the area, and it should be noted that the authors represent a total of 9 countries: Bangladesh, China, Hong Kong, India, New Zealand, Pakistan, Sweden, Turkey, and the United States.

As this is an edited book contributed by several authors from different geographical regions and different backgrounds, to keep the consistency of the subject matters discussed in each chapter, the book is organized in a sequence that puts the relatively easy-to-read papers in the beginning. Necessary modifications have been made in some chapters to reduce the potential redundancy or overlapping of discussed issues. However, some overlapping between chapters is in fact beneficial for the readers to connect different apparently disconnected but interrelated issues. I hope that the book will be a useful, especially for those researchers, graduate students, academicians, and practitioners who are working on the emerging area of crowd-assisted and crowd-associated networking and computing.

This is the first book of this kind on this topic, and hence, in the future, as the idea is further enhanced based on the advancements of different types of technologies, it is expected that future editions or other books in this area will be more comprehensive, covering issues that may be missing in this collection.

Al-Sakib Khan Pathan, PhD
Department of Computer Science and Engineering
Southeast University, Bangladesh
spathan@ieee.org, sakib.pathan@gmail.com

MATLAB® is a registered trademark of The MathWorks, Inc. For product information, please contact:

The MathWorks, Inc.
3 Apple Hill Drive
Natick, MA 01760-2098 USA
Tel: 508 647 7000
Fax: 508-647-7001
E-mail: info@mathworks.com
Web: www.mathworks.com

Acknowledgments

I sincerely thank the Almighty Allah for allowing me the time to complete another book. No such task is easy. It takes considerable time and sincere effort. Again, working in an environment like that in Dhaka, Bangladesh, is indeed difficult, where daily traffic jams kill many hours just on the road with other inherent chaos in daily life. I thank my wife, Labiba Mahmud, who always gives me the proper support and encouragement to carry on with my usual research activities. Again, both of my parents always encourage me to do more, especially my father, who often asks me about my next research plans and books. From the professional perspective, edited books are often quite difficult to put together as different authors have different views on the same topic. I must acknowledge that the authors have cooperated with me with their utmost sincerity. The reviewers also have done an excellent job. Last but not the least, I must thank Richard O'Hanley, who has always been supportive of me from the very beginning of the book's idea.

Editor

 **Al-Sakib Khan Pathan** earned his PhD degree in computer engineering in 2009 from Kyung Hee University, South Korea, and BSc degree in computer science and information technology from Islamic University of Technology, Bangladesh, in 2003. He is currently an associate professor in the Computer Science and Engineering Department, Southeast University (SEU), Bangladesh. He also holds the position of additional director of the Institutional Quality Assurance Cell, SEU. Previously, he was with the Computer Science Department at the International Islamic University Malaysia (IIUM), during 2010–2015 and with BRAC University, Bangladesh, during 2009–2010. He also worked as a researcher in the networking lab at Kyung Hee University, South Korea, from September 2005 to August 2009, where he completed his MS leading to his PhD. His research interests include wireless sensor networks, network security, cloud computing, and e-services technologies. Currently, he is working on some multidisciplinary issues. He is a recipient of several awards and best paper awards and has several notable publications in these areas. He has served as a general chair, organizing committee member, and technical program committee member of numerous international conferences and workshops, like INFOCOM, GLOBECOM, ICC, LCN, GreenCom, AINA, WCNC, HPCS, ICA3PP, IWCMC, VTC, HPCC, and SGIoT. He was awarded the Institute of Electrical and Electronics Engineers (IEEE) Outstanding Leadership Award for his role in the IEEE GreenCom'13 conference. He is currently serving as the editor-in-chief of the *International Journal of Computers and Applications* (Taylor & Francis); associate technical editor of *IEEE Communications Magazine*; editor of *Ad Hoc and Sensor Wireless Networks* (Old City Publishing), the *International Journal of Sensor Networks* (Inderscience Publishers), and the *International Arab Journal of Information Technology* (IAJIT); associate editor of the *International Journal of Computational Science and Engineering* (Inderscience Publishers); area editor of the *International Journal of Communication*

Networks and Information Security; and guest editor of many special issues of top-ranked journals. He has also been the editor or author of 15 published books. One of his books has been included twice in Intel Corporation's Recommended Reading List for Developers, second half 2013 and first half of 2014. Three books were included in the IEEE Communications Society's (IEEE ComSoc) Best Readings in Communications and Information Systems Security, 2013. Two other books were indexed with all the titles (chapters) in Elsevier's acclaimed abstract and citation database, Scopus, in February 2015, and a seventh book was translated to simplified Chinese language from the English version. Also, two of his journal papers and one conference paper were included under different categories in IEEE ComSoc's Best Reading Topics on Communications and Information Systems Security, 2013. He also serves as a referee for many prestigious journals. He has received some awards for his reviewing activities, such as an award for being one of the most active reviewers of IAJIT (several times) and the Elsevier Outstanding Reviewer of Elsevier *Computer Networks* (July 2015) and Elsevier *Journal of Network and Computer Applications* (November 2015). He is a senior member of the IEEE.

Contributors

Carlos Bermejo earned his MSc degree in telecommunication engineering in 2012 from Oviedo University, Spain. He is currently a PhD student at the Hong Kong University of Science and Technology, working in the System and Media Laboratory research group. His main research interests are Internet of Things, mobile augmented reality, network security, human–computer interaction, social networks, and device-to-device communication.

Md Zakirul Alam Bhuiyan is currently an assistant professor in the Department of Computer and Information Sciences at Fordham University, New York, New York. Earlier, he worked as an assistant professor at Temple University. His research focuses on dependability, cyber security, big data, and cyber-physical systems. He has served as a lead guest editor and associate editor for *IEEE Transactions on Big Data, ACM Transactions on Cyber-Physical Systems, Information Sciences, Future Generation Computer Systems, IEEE Internet of Things (IoT) Journal, Cluster Computing, The International Journal of Computers and Applications,* and so on. He has also served as a general chair, program chair, workshop chair, publicity chair, technical program committee member, and reviewer of international journals and conferences. He received the Institute of Electrical and Electronics Engineers (IEEE) TCSC Award for Excellence in Scalable Computing for Early Career Researchers (2016–2017) and the IEEE Outstanding Leadership Award (2016) and Service Award (2017). He is a senior member of IEEE and a member of the Association for Computing Machinery.

Bubu Bhuyan earned his MTech and PhD degrees from Tezpur University and Jadavpur University, respectively. He is currently serving as an associate professor in the Department of Information Technology, North-Eastern Hill University (NEHU), Shillong. His current area of research includes cryptographic algorithms, information theoretic security, and cloud security.

Metin Bilgin earned BSc and MSc degrees in computer systems and electronics from the University of Selçuk, Konya, Turkey, in 2005 and 2008, and a PhD degree in computer engineering from the University of Yıldız Technical, İstanbul, Turkey, in 2015. He is currently an assistant professor in the Department of Computer Engineering at Bursa Uludağ University, Bursa, Turkey. His research interests are text mining, data mining, natural language processing, machine learning (artificial intelligence in particular), mobile computing, dependency parsing, name entity recognition, and semantic role labeling.

Dimitris Chatzopoulos earned his diploma and MSc in computer engineering and communications from the Department of Electrical and Computer Engineering of the University of Thessaly, Volos, Greece. He is currently a PhD student in the Department of Computer Science and Engineering at the Hong Kong University of Science and Technology and a member of the System and Media Laboratory. His main research interests are in the areas of mobile augmented reality, mobile computing, device-to-device ecosystems, and cryptocurrencies.

Subhrajyoti Deb earned his MTech degree in computer science and engineering from the National Institute of Technology, Agartala, India, in 2015. He is currently a PhD scholar under the Visvesvaraya PhD Scheme in the Department of Information Technology, School of Technology, North-Eastern Hill University (NEHU), Shillong. He also worked as a visiting scientist at the Indian Statistical Institute, Kolkata. His research interests include cryptology, security issues in the Internet of Things, and steganography.

Sankhanil Dey is a registered PhD student of the A. K. Choudhury School of Information Technology, University of Calcutta, Kolkata, India. He earned his MTech from the Department of Radio Physics and Electronics, University of Calcutta, and was a senior research fellow at the Institute of Radio Physics and Electronics of the University of Calcutta until November 30, 2016. He was a teaching assistant as well as a guest teacher in the A. K. Choudhury School of Information Technology. His research interests include cryptography and cryptology, Boolean functions, applications of finite fields or Galois fields in cryptography, elliptic curve cryptography, lightweight cryptography, and the application of cryptographic algorithms in field-programmable gate array technology. He has authored many research articles in several archives, international journals, books, and conferences.

Marija Furdek is a researcher at the Optical Networks Lab at the KTH Royal Institute of Technology, Stockholm, Sweden. She earned her docent degree from KTH in 2017 and her Dipl-Ing (2008) and PhD (2012) degrees from the University of Zagreb, Croatia. Her research interests include optical and wireless network survivability, reliability analysis, physical layer security, and optimization techniques.

Marija has coauthored more than 70 publications in international journals and conferences, 4 of which received best paper awards. She serves as a general chair of the Photonic Networks and Devices Conference and as a guest editor of the *IEEE/OSA Journal of Optical Communications and Networking* and *IEEE/OSA Journal of Lightwave Technologies*.

Mario Gerla earned an engineering degree from the Politecnico di Milano, Italy, and a PhD degree from the University of California, Los Angeles (UCLA). He became an Institute of Electrical and Electronics Engineers (IEEE) fellow in 2002. As a graduate student at UCLA, he was part of the team that worked on the early Advanced Research Projects Agency Network system and protocols under the guidance of Professor Leonard Kleinrock. After 4 years at Network Analysis Corporation in New York, he joined UCLA's faculty in 1976. At UCLA he has designed network protocols, including ad hoc wireless clustering, multicast (ODMRP and CODECast), and Internet transport (TCP Westwood). He has led the ONR MINUTEMAN project, designing the next-generation scalable airborne Internet for tactical and homeland defense scenarios. He is now leading several advanced wireless network projects under industry and government funding. His team is developing a vehicular testbed for safe navigation, content distribution, urban sensing, and intelligent transport. His parallel research activities are wireless medical monitoring using smartphones and cognitive radio use in urban environments. He has served on several conference program committees, including MobiCom, MobiHoc, MedHocNet, and Wireless On-Demand Network Systems and Services (WONS). He is on the IEEE TON Scientific Advisory Board.

Ranjan Ghosh was born in November 1947 and is now associated with the Dumkal Institute of Engineering and Technology, Murshidabad, after complete retirement at the age of 65 from the Department of Radio Physics and Electronics of the Calcutta University as associate professor, having rendered continuous services since March 1980. Professor Ghosh earned BTech, MTech, and PhD (Tech) degrees at Calcutta University, from the same department, between 1967 and 1980. Between 1982 and 1984, he had a brief stay at the Ioffe Physico-Technical Institute, Leningrad (now St. Petersburg), for postdoctoral research. The area of his predoctoral and postdoctoral research interest was the simulation studies of IMPATT devices and their applications in microwave and optical engineering. Subsequently, he shifted to simulation studies of ion implantations undertaken in silicon process technology. His latest interest involves security issues of data, text, and video, covered by the field of cryptography.

Jairo Gutierrez is the deputy head of the School of Engineering, Computer and Mathematical Sciences at the Auckland University of Technology in New Zealand. He earned a systems and computing engineering degree from the Universidad de Los Andes in Colombia, a master's degree in computer science from Texas A&M University, and a PhD in information systems from the University of Auckland. His current research is on network management systems, networking security, viable business models for IT-enabled enterprises, next-generation networks, and cloud computing systems.

Pan Hui earned his PhD degree from the Computer Laboratory, University of Cambridge, and his MPhil and BEng both from the Department of Electrical and Electronic Engineering, University of Hong Kong. He is currently the Nokia chair in data science (with a generous endowment from Nokia) and a professor of computer science at the University of Helsinki. He has also been the director of the HKUST-DT System and Media Laboratory at the Computer Science and Engineering Department of Hong Kong University of Science and Technology since January 2013. He is an Association for Computing Machinery (ACM) distinguished scientist and an Institute of Electrical and Electronics Engineers (IEEE) fellow (Computer Society & Communications Society). He also serves as a distinguished scientist of Telekom Innovation Laboratories (T-labs), Germany and an adjunct professor of social computing and networking at Aalto University, Finland. Before returning to Hong Kong, he spent several years at T-labs and Intel Research, Cambridge. He has published more than 150 research papers and has some granted and pending European patents. He has founded and chaired several IEEE/ACM conferences and workshops and has served on the organizing and technical program committees of numerous international conferences and workshops, including ACM SIGCOMM, IEEE Infocom, ICNP, SECON, MASS, GLOBECOM, WCNC, ITC, ICWSM, and WWW.

Gunnar Karlsson has been a professor at the KTH Royal Institute of Technology since 1998, where he is the head of the Department of Network and Systems Engineering. He has previously worked for IBM Zurich Research Laboratory and the Swedish Institute of Computer Science. His PhD is from Columbia University, New York, and his MSc from Chalmers University of Technology in Gothenburg, Sweden. He has been a visiting professor at École Polytechnique Fédérale de Lausanne, Switzerland, and the Helsinki University of Technology in Finland and ETH Zurich in Switzerland. His current research relates to mobile communication, indoor localization, and quality of service.

Rahat Ali Khan earned his BS in electronics, MBA in finance, and MPhil in telecommunications from the University of Sindh, Jamshoro, Pakistan, in 2005, 2012, and 2016, respectively. He is currently serving as assistant professor and pursuing his PhD in information technology at the Institute of Information and Communication Technology, University of Sindh, Jamshoro, Pakistan. He has been working as a research associate/teaching assistant at the Institute of Information and Communication Technology since 2007. His research interests include wireless sensor networks, wireless body area sensor networks, underwater acoustic sensor networks, mobile crowdsensing, cyber-physical systems, and Internet of Things. He serves as a reviewer for recognized journals of the Institute of Electrical and Electronics Engineers, SAGE Publications, and Taylor & Francis, and for the Institution of Engineering and Technology's *Electronics Letters*. He received an exceptional reviewer certificate from *Electronics Letters*.

Xue Jun Li earned his BEng (first class honors) and PhD in electrical and electronic engineering from Nanyang Technological University (NTU), Singapore, in 2004 and 2008, respectively. Between November 2007 and July 2008, he worked as a research engineer and then as a research fellow at the Network Technology Research Centre, NTU. Between August 2008 and September 2008, he worked as a research scientist at Temasek Laboratories, NTU. From September 2008 to May 2011, he was with the School of Electrical and Electronic Engineering, NTU, as a faculty member. Between June 2011 and January 2013, he worked as a research scientist at the Institute for Infocomm Research, Agency for Science, Technology and Research, Singapore. Since January 2013, he has been with the School of Engineering, Auckland University of Technology, where he is a senior lecturer. His research interests include design and analysis of wireless networking protocols, modeling and design of radio frequency integrate circuits, and system optimizations.

 **Yang Li** earned her BSc degree in computer science from China Three Gorges University in 2015. Currently, she is a graduate student at the National Huaqiao University. Her research interests include wireless networks, Internet of Things, and mobile computing.

 **William Liu** is currently a senior lecturer in the Department of Information Technology and Software Engineering, School of Engineering, Computer and Mathematical Sciences at the Auckland University of Technology, New Zealand. He holds a master's degree and PhD degree in electrical and computer engineering, both obtained at the University of Canterbury, New Zealand. He had been working as a network planner and designer at Beijing Telecom for 5 years before he immigrated to New Zealand. He has co-authored more than 75 papers published in international journals and conferences, and he participates in the program committees of several premier Institute of Electrical and Electronics Engineers (IEEE) conferences. His main research interests are in the design and performance evaluation of the next-generation network architecture and protocols. He is working especially in the areas of network survivability, sustainability, and trustworthy computing.

 **Hao Luo** earned his BS degree from the Civil Aviation University of China in 2016. Currently, he is a master candidate at the National Huaqiao University of China. His research interests include wireless sensor networks, mobile computing, and fog computing.

Rashmi Munjal is a current PhD student in the Department of Computer and Mathematical Sciences at Auckland University of Technology in New Zealand. She earned her master's degree (honors) in computer sciences from Kurukshetra University in 2013. She completed her bachelor of technology in information technology from Kurukshetra University in 2008. Her research interests are green networking, big data dissemination, vehicular networks, and delay-tolerant networking

İzzet Fatih Şentürk earned a bachelor's degree in computer engineering from Ege University in 2006 and master's and PhD degrees in computer science from Cornell University and Southern Illinois University–Carbondale in 2008 and 2013, respectively. He is currently an assistant professor in the Department of Computer Engineering at Bursa Technical University. He is a member of Institute of Electrical and Electronics Engineers. His research interests include wireless sensor networks, mobile computing, pervasive computing, and cloud computing.

Avinash Srinivasan has held numerous academic positions and other affiliations since 2008. Currently, he is an associate professor in the Computer and Information Sciences Department at Temple University (TU) and a fellow of the National Cybersecurity Institute in Washington, D.C. Prior to joining TU in the summer of 2014, he was an assistant professor in the Computer Science Department at George Mason University from spring 2012 to spring 2014 and an assistant professor of computer forensics at Bloomsburg University from fall 2008 to fall 2011. He is also a certified ethical hacker and computer hacking forensics investigator. Dr. Srinivasan's research interests broadly span the areas of cybersecurity and digital forensics. He has published 47 refereed papers in scholarly conferences and journals, including IEEE INFOCOM, ACM SAC, IEEE ICC, IEEE ICDCS, and IEEE MALWARE, and was the recipient of the best paper award at ICITST 2012. His research work publications have been cited 850 times. Since 2008, Dr. Srinivasan has been involved as the principal investigator or co-principal investigator on federally funded research from agencies including the Department of Education, Department of Justice, Department of Homeland Security, National Science Foundation, and Department of Defense/Navy. Dr. Srinivasan has more than 450 hours of formal training in cybersecurity and digital forensics. Since 2008, Dr. Srinivasan has trained law enforcement officers and civilians in various cybersecurity and digital forensics topics.

Sabu M. Thampi is an associate professor at the Indian Institute of Information Technology and Management–Kerala, Trivandrum, India. He completed his PhD in computer engineering from the National Institute of Technology, Karnataka. His research interests include network security, security informatics, bio-inspired computing, video surveillance, cloud security, secure information sharing, secure localization, and distributed computing. He has authored and edited a few books published by reputed international publishers and published papers in academic journals and international and national proceedings. He is currently serving as an editor for the *Journal of Network and Computer Applications*, Elsevier, and the *Journal of Applied Soft Computing*, Elsevier; an associate editor for *IEEE Access* and *International Journal of Embedded Systems*, Inderscience, UK; and a reviewer for several reputed international journals. He is a senior member of the Institute of Electrical and Electronics Engineers (IEEE) and a member of the IEEE Communications Society; IEEE Systems, Man, and Cybernetics Society; and Association for Computing Machinery.

Rohit Upadhya is currently pursuing his BTech in computer science and engineering from the National Institute of Technology, Silchar, Assam. He completed his secondary education from St. Edmund's School, Shillong, and his senior secondary education from South Point School, Guwahati. His areas of interest include cryptography, big data, and competitive coding.

Elizabeth B. Varghese is a doctoral research scholar at the Indian Institute of Information Technology and Management–Kerala, India. She earned her MTech degree in computer science, with a specialization in digital image computing, from the Department of Computer Science, University of Kerala, in 2011 and BTech degree in information technology from the Cochin University of Science and Technology in 2003. Her research interests include crowd analysis, video analytics, data analytics, and Internet of Things.

Tian Wang earned his BSc and MSc degrees in computer science from Central South University in 2004 and 2007, respectively. He earned his PhD degree from City University of Hong Kong in 2011. Currently, he is a professor at Huaqiao University, China. His research interests include wireless sensor networks, fog computing, and mobile computing.

Jiyuan Zhou earned his BS degree from Tianjin Polytechnic University in 2016. Currently, he is a master candidate at Huaqiao University, China. His research interests include security in wireless networks, fog computing, and security in cloud storage.

Contributor List

Carlos Bermejo
Hong Kong University of Science and
    Technology
Hong Kong

Md Zakirul Alam Bhuiyan
Fordham University
New York, New York

Bubu Bhuyan
North-Eastern Hill University
Shillong, India

Metin Bilgin
Bursa Technical University
Bursa, Turkey

Dimitris Chatzopoulos
Hong Kong University of Science
    and Technology
Hong Kong

Subhrajyoti Deb
North-Eastern Hill University
Shillong, India

Sankhanil Dey
University of Calcutta
Kolkata, India

Marija Furdek
KTH Royal Institute of Technology
Stockholm, Sweden

Mario Gerla
University of California at Los Angeles
Los Angeles, California

Ranjan Ghosh
University of Calcutta
Kolkata, India

Jairo Gutierrez
Auckland University of Technology
Auckland, New Zealand

Pan Hui
Hong Kong University of Science
    and Technology
Hong Kong

Gunnar Karlsson
KTH Royal Institute of Technology
Stockholm, Sweden

Rahat Ali Khan
University of Sindh
Jamshoro, Pakistan

Xue Jun Li
Auckland University of Technology
Auckland, New Zealand

Yang Li
Huaqiao University
Xiamen, China

William Liu
Auckland University of Technology
Auckland, New Zealand

Hao Luo
Huaqiao University
Xiamen, China

Rashmi Munjal
Auckland University of Technology
Auckland, New Zealand

Al-Sakib Khan Pathan
Southeast University
Dhaka, Bangladesh

İzzet Fatih Şentürk
Bursa Technical University
Bursa, Turkey

Avinash Srinivasan
Temple University
Philadelphia, Pennsylvania

Sabu M. Thampi
Indian Institute of Information
    Technology and
    Management–Kerala
Trivandrum, India

Rohit Upadhya
National Institute of Technology
    Silchar
Assam, India

Elizabeth B. Varghese
Indian Institute of Information
    Technology and
    Management–Kerala
Trivandrum, India

Tian Wang
College of Computer Science and
    Technology
Huaqiao University
Xiamen, China

Jiyuan Zhou
Huaqiao University
Xiamen, China

Chapter 1

An Introduction to Mobile Crowdsensing

Rahat Ali Khan and Al-Sakib Khan Pathan

Contents

1.1 Introduction

One of the fundamental goals of many research works is to add value and ease to human life and, eventually, to benefit humankind [1–3]. The tremendous development and advanced research activities in the fields of information and communication technology (ICT) in the past two decades have led to the realization of some significant technological aspects of life that were never realized before [4]. Like ICT, the fields related to electronics and circuit technologies have also obtained remarkable cost reduction, alongside a significant level of sophistication. Due to these facts, we have witnessed rapid growth in the use of mobile devices in the past decade, especially smartphones in recent years. According to a report presented in [4], there will be 4.93 billion mobile phone subscribers by the end of 2018. The number of mobile phone subscribers statistics from 2013 to 2019 (expected) are shown in Figure 1.1 [5].

Nowadays, many cell phones have sensors embedded into them, alongside significant computing capabilities. Hence, with their multifunctional capabilities, these devices could be considered some kind of computer. While just a decade ago cell phones were mainly for voice communication, today their advanced versions, called *smartphones*, have become an important part of information sharing, enabling the progress of mobile sensing technologies [6].

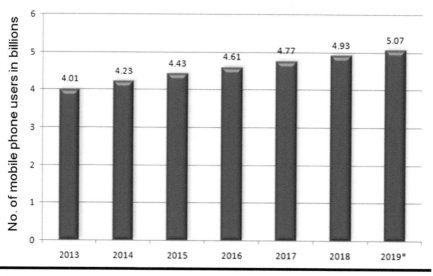

Figure 1.1   Mobile phone subscribers over the years (approximately and expected).

Many of today's smartphones are able to sense the surrounding environment and transmit recorded data for sharing and processing purposes. For example, smartphones can easily capture information regarding movements of objects and noise levels. Parameters like pollen level in the air, humidity, temperature, and pollution can also be captured with these smartphones if some special-purpose sensors are embedded into them. The captured information can be used to plan routine activities, helping the users schedule their routine life to avoid environmental conditions that could risk their lives [7]. In fact, there are other specialized circuits and sensors that can make smartphones capable of sensing location with the help of the Global Positioning System (GPS), measuring the speed of traveling or movement with the aid of an accelerometer, measuring noise level with the help of a microphone, and so on. Some common embedded devices used today with smartphones are a digital compass, light sensor, GPS, and camera [8]. Hence, it is now possible even, for instance, to detect the locations of potholes with the help of these technologies.

Arguably, the term *mobile crowdsensing* was first formally coined in [9]. Mobile crowdsensing is often called simply *crowdsensing*. However, we argue that in some scenarios, mobility may be weakly involved and the crowd could still generate data and information. For instance, different people report an event from immobile devices from different locations. Hence, crowdsensing and mobile crowdsensing have slightly different connotations in some scenarios. Various crowdsourcing techniques are used to manage the entire mobile sensing ecosystem. Together with static sensor networks and the Internet of Things (IoT), mobile crowdsensing has the potential to make the vision of *ubiquitous sensing* a reality.

So far, we have used the term *crowd*, but a question arises here: "What is this crowd composed of?" The crowd is basically a set of human beings who carry devices that generate data that could be gathered and shared to make sense out of some incident. While people can generate the data, the sensors embedded into the mobile devices and other static devices can also generate the data. In simple terms, the humans in this setting become the citizens of the sensing environment [10].

In Figure 1.2, the mobile sensing components are shown, which are sensors in smartphones, applications in smartphones, and application programming interfaces (APIs). Given the current trend, it is expected that more types of sensors or sensing technologies would be embedded into future smartphones. Hence, the capabilities of those devices would increase manifold. In any case, humans would be the key participants in such environments who would allow data of common interest to be shared and accessed via their mobile and other types of devices [11].

1.2 Sensors in Smartphones

Different types of sensors can be embedded into a smartphone. In this section, we mention some of them that can be used for crowdsensing. We will explore their functionalities and try to understand how these sensors can really be beneficial in

MOBILE SENSING

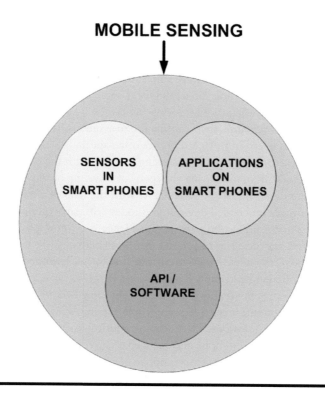

Figure 1.2   Components of mobile phone sensing.

such a context. Figure 1.3 shows a diagram that summarizes the sensing capabilities of an advanced smartphone today.

Compass—Smartphones have a built-in magnetometer sensor on which the compass is based. It provides the north direction depending on the location of the phone by the help of the magnetic field of the earth.

Proximity—The proximity sensor is placed near the earpiece of the mobile phone. It consists of an infrared (IR) light detector and an IR light-emitting diode (LED). When a mobile phone is placed near the ear, the proximity sensor recognizes and informs the phone that there is or may be a call in process and so turns off the screen to save battery. This detection is performed by IR light, which the proximity sensor throws outwards and is reflected back to the human object.

Gyroscope—This is basically linked with the accelerometer. It detects the rotation of the phone to enhance a game experience.

Ambient light—Ambient light sensors are used in mobile phones for the purpose of adjusting the brightness automatically. They react to adjust the brightness depending on the surrounding light.

Accelerometer—The accelerometer is basically used by applications in mobile phones for detecting the orientation of the mobile phone, either portrait or

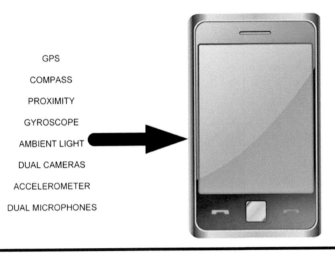

Figure 1.3 **Sensors available in modern smartphones.**

landscape. It also measures tilting motion. Depending on how the phone is being held, it responds and presents the users with a correct and suitable view of photographs, videos, or any text files.

Fingerprint sensors—As per the need today, smartphones can now have built-in fingerprint sensors embedded into them. The main application of these sensors is to sense the fingerprint pattern and, based on it, lock or unlock the phone. In some mobile phones, there is an option to install an application that lets the camera interact with the fingerprint sensor to take pictures [12].

Pedometer—This is a sensor that is related to health issues. It counts the number of steps. If a mobile phone is kept in one's pocket and the user walks or runs, the pedometer interacts with its movements and, based on them, senses and counts the steps.

The sensors in smartphones are classified into four categories:

1. Communication
2. Position and motion
3. Touch, camera
4. Environmental

Table 1.1 shows the categories of the sensors available in today's smartphones.

1.3 Types of Mobile Crowdsensing

Now we turn to the types of mobile crowdsensing. Mobile crowdsensing is sometimes also called *people-centric sensing*. *People-centric sensing* is the term used when various matters of a person are considered, such as *likes* and *dislikes*, what he is up

Table 1.1   Sensors in Smartphones

Sensor Category	Sensors
Communication	Global System for Mobile (GSM) Bluetooth Global Positioning System (GPS) Wireless Fidelity (Wi-Fi)
Position and motion	Accelerometer Pedometer Proximity sensor Magnetometer Gyroscope
Touch, camera	Touch sensor Microphone Fingerprint Camera Camera flash
Environmental	Temperature sensor Barometer Air humidity sensor Radiation sensor Ambient light sensor

to, his interests, where he goes, and other personal traits. People-centric sensing is the combination of crowdsensing and IoT, which provides a platform to users to share sensed data and events in their surroundings, as well as ideas [13].

As there are various scenarios that could be similar to mobile crowdsensing, we broadly classify crowdsensing into the following categories [9, 14] (Figure 1.4):

1. Participatory sensing—Smartphone/smart device performs sensing.
2. Opportunistic sensing—Manual involvement of the participant is needed.

Again, when the person is the main focus, people-centric sensing is categorized as

1. Personal sensing
2. Social sensing
3. Public sensing

Personal sensing—This type of sensing focuses on personal monitoring. Personal sensing systems monitor or share the custodian's personal information, that is, the information that the custodian of the sensing device deems to be sensitive. Such systems collect the custodian's information, for instance, his or

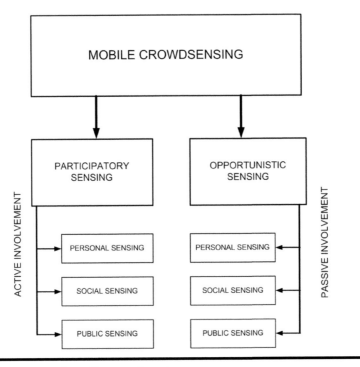

Figure 1.4 Types of mobile crowdsensing.

her daily life patterns, physical activities, health readings (e.g., heart rate, blood pressure, and sugar level), personal and social contacts, and location, for further processing.

Social sensing—In this type of sensing, each individual plays the role of a sensor and generates or shares data. Data could be acquired from different social networking profiles. The proliferation of the use of Facebook, Twitter, and similar platforms has enabled this kind of sensing, where the social network can be used to collect data about various individuals, systems, policies, and so forth.

Public sensing—The basic idea of public sensing is the automatic collection of sensor data from mobile devices that are carried and owned by the general public or commoners. This mechanism enables low-cost provisioning of real-time sensor data considering spatial and environmental dimensions, which can be used for a variety of applications.

1.3.1 Participatory Sensing

In this type of sensing, users are directly involved in the process of sensing. It is also called manual sensing. The individuals involved perform actions like taking pictures or recording noise levels or their positions for contributing to the sensing operation. It can be said that in participatory sensing, users, that is, the people carrying mobile phones, are acting as nodes and form a sensor/sensing network with other devices nearby. People carry smart watches, laptops, personal digital assistants (PDAs), and GPS receivers, and even have vehicles equipped with sensors. The mobile phone is the most common device for ordinary people to collect and then transfer data because such a setting does not require any preinstalled infrastructure [15, 16].

1.3.1.1 Participatory Personal Sensing

Some of the proposed participatory personal sensing systems are listed in Table 1.2.

1.3.1.2 Participatory Social Sensing

Some of the proposed participatory social sensing systems are listed in Table 1.3.

1.3.1.3 Participatory Public Sensing

Some of the proposed participatory public sensing systems are listed in Table 1.4.

1.3.2 Opportunistic Sensing

In opportunistic sensing, the user has relatively less involvement in the sensing operation. Hence, this kind of sensing method is sometimes called *automatic sensing*. What happens in this case is that the smartphone itself makes the sensing-related decisions when the related sensor is activated. Such automatically sensed data are often transmitted to the concerned server [32].

Table 1.2 Participatory Personal Sensing Systems

S. No.	Proposed System	Proposed by	Sensors
1	HealthAware	[17]	GPS Accelerometer Camera
2	Hyper Fit	[18]	Camera
3	CONSORTS-S	[19]	Camera
4	BALANCE	[20]	GPS Accelerometer
5	UbiFit Garden	[21]	Accelerometer
6	Mobi Care Cardio	[22]	Accelerometer Wireless ECG

Table 1.3   Participatory Social Sensing Systems

S. No.	Proposed System	Proposed by	Sensors
1	DietSense	[23]	Microphone GPS Camera
2	MoVi	[24]	Microphone Accelerometer Camera
3	CenceMe	[25]	Camera Accelerometer Microphone GPS Radio 3 BlueCel Sensor
4	Triple Beat	[26]	Two-lead ECG sensor Triaxial accelerometer

Table 1.4   Participatory Public Sensing Systems

S. No.	Proposed System	Proposed by	Sensors
1	PetrolWatch	[27]	Camera GPS sensor
2	TrafficSense	[28]	Phones Microphone Camera GPS sensor Accelerometer GSM radio
3	Citizen Journalist	[29]	Phones Microphone Camera GPS sensor Accelerometer
4	NoiseTube	[30]	Microphone GPS sensor
5	V-Track	[31]	GPS sensor

1.3.2.1 Opportunistic Personal Sensing

Some of the proposed opportunistic personal sensing systems are listed in Table 1.5.

1.3.2.2 Opportunistic Social Sensing

Some of the proposed opportunistic social sensing systems are listed in Table 1.6.

1.3.2.3 Public Opportunistic Sensing

Some of the proposed opportunistic public sensing systems are listed in Table 1.7.

1.4 Mobile Crowdsensing Applications

As already understood, mobile crowdsensing can be very useful in many different kinds of applications. In general, the available applications of mobile crowdsensing today are classified as three [9] main types:

1. Environmental monitoring applications
2. Infrastructure monitoring applications
3. Social monitoring applications

In the subsequent paragraphs, each of these types is described.

1.4.1 Environmental Applications of Mobile Crowdsensing

In this case, the target area is the natural environment, for example, the tasks related to wildlife habitat monitoring, the level of water in creeks, or a city's pollution level. Already, there are a few implemented applications for these, as stated below.

Common Sense—A prototype has been developed by the researchers in [41], which is known as Common Sense. The prototype is focused on the detection of

Table 1.5 Opportunistic Personal Sensing Systems

S. No.	Proposed System	Proposed by	Sensors
1	Emotion Sense	[33]	Accelerometer Microphone
2	Heart Gear	[34]	Oximetry sensor
3	Heart ToGo	[35]	ECG sensor Triaxial accelerometer
4	PerFall ID	[36]	Accelerometer
5	I-Fall	[37]	Triaxial accelerometer GSP sensor

Table 1.6 Opportunistic Social Sensing Systems

S. No.	Proposed System	Proposed by	Sensing Mechanism
1	WhozThat	[38]	Social networking
2	Opportunistic Localization System (OLS)	[39]	Localization

Table 1.7 Opportunistic Public Sensing Systems

S. No.	Proposed System	Reference	Sensors
1	Road Bump Monitor	[29]	Microphone Camera GPS sensor Accelerometer
2	NeriCell	[40]	Microphone GSM radio GPS sensor Accelerometer Camera
3	Mobile Century	[50]	GPS sensor

pollutants in the air by using some specialized sensors. According to their proposed prototype, the mobile phone captures the data and sends it after necessary processing. The recorded data can be transferred through any of the various technologies, like IEEE 802.15.4 or Bluetooth. The collection of the mobile phone's data can be sent through wireless sensor networks (WSNs). A WSN provides sensor infrastructure in the presence of devices and fills the gaps for those areas where there is no infrastructure [3].

Creek Watch—The Creek Watch project was developed by IBM researchers at the Center for Social Business, especially the Almaden Lab [42]. This is basically an application for the iPhone. It is a type of crowdsensing in which the objective is to locate and monitor the health of watersheds. When a group of people plan to visit an area and they find a watershed there, they are only required to spend a few moments using the Creek Watch app, which will take pictures of how much water (level) is there or measure how polluted the water is. These data then could be sent to an IBM (company) office. IBM aggregates the data and communicates with officials so that they can manage the water resources and take proper steps toward removing or reducing pollution. Creek Watch monitors the watershed and pollution level as

1. Amount of water
2. Flow rate of water

3. Amount of trash
4. Snapshots of the waterway

Personal Environmental Impact Report (PEIR)—This was developed by the authors in [32]. PEIR was developed with the aim of nature conservation, air pollution monitoring, environment monitoring, and monitoring of other parameters that may have an impact on the environment. PEIR uses data of different locations gathered from mobile phones for the purpose of measuring environmental conditions and their overall impact. This is based on data collected from GPS embedded into mobile phones that are sent to a server. The server gathers the data and processes it in stages as hidden Markov models (HMMs) for the purpose of determining the mode of transportation and data segmentation.

NoiseTube—Introduced by the researchers in [32], this application was developed in order to assess noise pollution with the help of local residents. With this application, users or participants detect various noise levels at the places they visit or travel to. Then this information is added to other information provided by the user, like which location it is being recorded from or, perhaps, some other annotations, and then the information is shared with others. Using this app, all other participants do the same, and in this way, there is a comprehensive noise map that is collected and stored at the server side for the purpose of analysis and future use.

1.4.2 Infrastructure Applications of Mobile Crowdsensing

Mobile crowdsensing can also be very useful when it comes to collecting data for public infrastructure. In fact, this can be done at a very low cost using mobile crowdsensing. Users having smartphones can collect data to help government institutions overcome traffic congestion problems. They can report several issues related to traffic, like the road conditions, availability/unavailability of parking lots, potholes in the road, and malfunctioning street lights.

NeriCell [40], developed by Microsoft Research, and CarTel [43], developed by MIT, are two example systems being used for the detection of road traffic–related problems. Other examples are noted below.

ParkNet—The researchers in [55] present an application that is for drivers. This app informs them about the availability of parking space. It uses a vehicular sensing system that runs over a mobile ad hoc sensor network (MANET) that consists of vehicles. This system basically collects data about available parking spaces and broadcasts this information in real time in urban areas.

Traffic Sense—This app is proposed in [28]. It has been designed to offer assistance in monitoring road bumps, traffic jams, potholes, and emergency situations. It is helpful in improving transportation infrastructure. The operation of this app is based on the contribution of participants using their smartphones in locations where they feel there is a need for improvement.

PetrolWatch—Proposed by [27], this is another application that is today possible due to the concept of crowdsensing. The prices of fuel are collected by this application. The goal of developing this is to inform drivers or the general population about the current status of fuel prices, especially information about where they can get fuel at a relatively lower price. There are several algorithms used for this application as it uses a camera to get an image of the prices displayed at fuel pump stations. The data are saved for later use and for analyzing fuel price trends.

1.4.3 Social Applications of Mobile Crowdsensing

Mobile crowdsensing enables interesting and useful social applications. For instance, data of exercise and workouts can be shared with others in a community. In this section, we look at a few works in this specific area.

Bike Net—The researcher in [43] has presented a work named Bike Net, which is a sensing system for cyclists. It has various features that may help cyclists enhance their rides, like selecting a clear route instead of a bumpy track and finding their destinations' distances. This is all done with the help of phone users available in a certain area. Bike Net provides support to cyclists as well as to the community through social networking using the Bike View Portal.

Party Thermometer—This is proposed in [29] and is intended for parties and party lovers. Several queries can be asked to partygoers during a party about their experience, and their feedback is collected and can then be sent to others to interest them in joining the party. This application uses several parameters to detect whether an individual is actually present at the party (and not just a passerby). The mobile phone must be stable (static) once it is at the party, and the app should work on the user's mobile system. The application uses fast Fourier transform (FFT) to detect music signals and frequency spikes. This is done to make sure that the user is at a party (i.e., actively staying/participating there). For this function to be performed, the microphone of the smartphone is utilized (which is kept turned on for capturing music signals). The application is designed in a way to make sure that the battery of the smartphone is not consumed much. First, it detects the location of the person, and only after making sure that the smartphone is in the building or place where the party is under way does the microphone turn on for further processing.

1.4.4 Mobile Crowdsensing and Wireless Sensor Networks

A WSN is a type of wireless network [51]. In a WSN, the tiny sensor nodes cumulatively monitor changes in an environment. After monitoring for some period, readings (like temperature, light intensity, magnetism, seismic activity, and movement—any of these or multiple in parallel) are sent to a more powerful node called a *sink* or *base station* for further processing. WSNs have important applications in remote environmental sensing, natural disaster relief, exploration of hazardous environments, and so on. The sensors used in a WSN could be microelectromechanical system (MEMS)–based sensors or sensors that are embedded into the smartphones. What makes WSN a suitable and relatively better candidate in environment monitoring is its ad hoc network infrastructure. An ad hoc network infrastructure is preferable for use in areas where there is no preplanned network deployment. Hence, when the nodes are active, such a network could perform monitoring operations [44].

A WSN-like network can be formed by using the concept of mobile crowdsensing. For such an ad hoc network, we only need to have a group of people gathered at some point in close proximity who have sensor-carrying smartphones. All these sensors can connect with some mechanism (if the users allow), and that collective set of sensors, like (forming) a WSN, could take readings of various parameters based on their capabilities. Such a mobile phone– or smartphone-based WSN can be used in various different monitoring scenarios for surveillance, intrusion detection, infrastructure protection, and so on [45–46], [51].

In a traditional WSN, the sensors are small, but more importantly, their energy resources are very limited. Irrespective of the reality the sophistication of technology takes us to, it is expected that the tiny devices (whatever they are called in the future), with very low energy resources, will also be important for some applications in the future. In the traditional WSN, once the network is deployed there is very little scope of recharging the sensor batteries; smartphone-based sensors have an advantage in this respect. A smartphone can be recharged when needed, and nowadays we also have power banks that can be carried in the pocket! Though various research works are being done on the energy efficiency issues in WSNs [47] and on energy-harvesting techniques [52], those are mainly intended for use by the separate sensor nodes. In fact, the embedded sensors in the smartphones may not need the energy-harvesting techniques for them because the users of the smartphones will often recharge their devices using direct power cables (i.e., embedded sensors will automatically get a regular power source [charged] when the smartphone is charged). Hence, with such a combination of smartphone technology and WSN-like operation, we can obtain better crowdsensing platforms in the coming days.

1.4.5 Mobile Crowdsensing and Surveillance

Mobile crowdsensing can play a vital role in surveillance. This section discusses some of the works that are based on surveillance systems using mobile crowdsensing.

Almost every person has been affected by some kind of traffic event in life, for instance, getting stuck in a traffic jam when there is an emergency situation (or when you need to reach a destination as soon as possible), traffic accident, or any other traffic mishap, like a bridge collapse or a broken wheel of a vehicle causing huge traffic jam on all sides. In such cases, crowdsensed data [50] could be used to warn other vehicles to avoid a particular road or to find alternative routes or means to reach the destination. In fact, nowadays *Google Maps* has a live traffic feature that shows congested portions of a road; it uses the data collected from hundreds and thousands of mobile phones. Such use of crowdsensing with real-time surveillance definitely improves the quality of life and solves some urban environmental and intracity problems. Another interesting application is when thousands of people gather for some event, for instance, a religious gathering like Hajj. Sometimes, we hear about cases of stampedes and deaths and injuries. One of the saddest and major crowd disasters happened in recent times in Mecca, where people gathered to fulfill their Hajj obligation. It was reported that about 1100 people lost their lives and approximately 900 were injured [48]. Such incidents could be prevented by real-time crowdsensing-based surveillance so that the pilgrims could get the right information about where to go or where to wait until a heavy-crowd situation eases.

1.4.6 Mobile Crowdsensing in Healthcare Sector

Mobile crowdsensing can be utilized to monitor various diseases in humans. Data regarding human health in their daily routines (checking glucose level for diabetic patients), health parameters (heart rate, pulse rate, and blood pressure), and physical activities (walking, jogging, and exercise) are collected by the sensing devices for ubiquitous monitoring [49]. In fact, there is a prototype system named *DietSense* [23] that takes pictures of meals and shares them with the community. This is helpful to those who are diet conscious. As the pictures are shared, the user gets suggestions from others and then may modify his or her meal routine to keep healthy. This can also be helpful for patients to get suggestions about what to take and what to avoid during an illness.

1.5 Challenges

While mobile crowdsensing is very promising as a concept and there are already several applications available based on it, there are still some challenges that exist that may restrict its future growth and its becoming more effective in regular human life.

1.5.1 Use of Extra Sensors

There are a limited number of sensors that are available in mobile phones. In fact, with the circuitry of the mobile phone, only a few sensors can be attached or embedded. Even if future technologies become more advanced, the size and platform of the mobile phone would remain the deciding factor for how many or which types of sensors could be placed on it. If in a particular case it is necessary to have some different types of sensors than those that are available in a mobile phone, then the need may arise to use external sensors in a crowdsensing application or scenario. This kind of external sensor would need to be connected to the mobile phone in a wired or wireless way, but such a connection would cause the mobile phone's battery to drain more rapidly. Another problem is the placement of such sensors because they would occupy space since they are not embedded into the phones. Hence, as a requirement, such external sensors should be lightweight and significantly small in size to make them suitable for carrying. Also, they should be very low-power-consuming devices and not consume much energy from the mobile phone's battery or have some other advanced battery source. This requires more improvement in battery technologies in the coming days.

1.5.2 Coverage

Mobile crowdsensing could be compared with the event coverage operation in traditional WSNs. For successful deployment and design of mobile crowdsensing systems, factors like number of participants and their relationships with the event coverage need to be considered. Issues include the spatial distribution of the participants, the capability of their mobile phones, the remaining battery power for mobile phones, and the disconnected segment of the ad hoc mobile network.

1.5.3 Grouping and Interactions

In many of the existing mobile crowdsensing systems, one thing that is often lacking is interaction between participants. Group interaction is a very important factor because this enhances the quality of the data in the system. If interaction is made a necessary condition of mobile crowdsensing, there exists the challenge of how the participants will be grouped together and how they will or should interact with each other.

1.5.4 Security and Privacy

Security and privacy are two very critical challenges and concerns for any type of crowdsensing. Because often smartphones/mobile phones carried by users need to be identified and the location information needs to be accessed for data collection, many users may not feel comfortable participating in such actions. In fact, many users turn their location information option off for better privacy as one may not want to share data with the community. There also remain various issues of potential security breaches [53,54] if certain types of access to the mobile phone are given to others who a user does not know or fully trust.

1.6 Final Words

As the trend of using smartphones has increased rapidly all over the globe, even in developing countries, it has led to increased use of the Internet. In fact, the massive use of the Internet has made people more pervasive in the sense that they are online almost 24/7. This situation has basically given birth to technologies like mobile crowdsensing, IoT, and cyber-physical systems (CPSs). As more and more people are becoming addicted to the Internet, even sometimes for not so important online activities, they are becoming more useful in contributing to the operations of mobile crowdsensing.

The advantage that mobile crowdsensing offers is that instead of deploying or installing a conventional and fixed infrastructure-based data collection network, the crowd's mobile phones can be used with relatively lower cost and faster deployment mechanisms. This kind of sensing can really be helpful in getting the readings of different parameters in remote areas whenever the users visit there in a random pattern (e.g., tourists). This is indeed a very effective technology for various situations as often the users just move around and need to contribute very little. Just the use of their mobile phones (to perform the tasks of data collection) helps a lot in making better sense of a certain region's temperature, rain, traffic jams, water levels, volcanic activities, and so on. In fact, this could be a significant way of gaining critical data that could be used by governments and organizations for their future plans of improving the quality of life of a certain region, taking necessary measures for emergency rescue operations or controlling violent crowds, dealing with political chaos, helping homeland security, and so on.

Acronyms

API	Application programming interface
FFT	Fast Fourier transform
GPS	Global Positioning System
GSM	Global System for Mobile
HMM	Hidden Markov model
ICT	Information and communication technology
IoT	Internet of Things
IR	Infrared
LED	Light-emitting diode
MEMS	Microelectromechanical system
OLS	Opportunistic localization system
PDA	Personal digital assistant
Wi-Fi	Wireless Fidelity
WSN	Wireless sensor network

References

1. R A Khan, S Memon, J H Awan, H Zafar, and K H Mohammadani, "Enhancement of Transmission Efficiency in Wireless On-Body Medical Sensors," *Engineering Science and Technology International Research Journal*, vol. 1, no. 2, pp. 16–21, 2017.
2. R A Khan, S Memon, S Zardari, L D Domeja, and Muhammad Usman, "Transposition Technique for Minimization of Path Loss in Wireless On-Body Medical Sensors," *Sindh University Research Journal (Science Series)*, vol. 48, no. 4, pp. 747–754, 2017.
3. Rahat Ali Khan et al., "An Energy Efficient Routing Protocol for Wireless Body Area Sensor Networks," *Wireless Personal Communications*, doi: 10.1007/s11277-018-5285-5, 2018.
4. I H Jaffri, S Soomro, and R A Khan, "ICT in Distance Education: Improving Literacy in the Province of Sindh Pakistan," *Sindh University Journal of Education*, vol. 40, no. 2010–11, pp. 37–48, 2011.
5. Statistica, Number of Mobile Phone Users Worldwide from 2013 to 2019 (in billions) [Online]. https://www.statista.com.
6. X Zhang et al., "Incentives for Mobile Crowd Sensing: A Survey," *IEEE Communications Surveys & Tutorials*, vol. 18, no. 1, pp. 54–67, 2016.
7. L G Jaimes, I J Vergara-Laurens, and A Raij, "A Survey of Incentive Techniques for Mobile Crowd Sensing," *IEEE Internet of Things Journal*, vol. 8, no. 5, pp. 370–380, 2015.
8. S Mardenfeld et al., "GDC: Group Discovery Using Co-Location Traces," in *Social Computing (SocialCom), 2010 IEEE Second International Conference on*, Minneapolis, MN, 2010.
9. R K Ganti, Fan Ye, and Hui Lei, "Mobile Crowdsensing: Current State and Future Challenges," *IEEE Communications Magazine*, vol. 49, no. 11, pp. 32–39, 2011.
10. K Farkas and I Lendak, "Simulation Environment for Investigating Crowd-Sensing Based Urban Parking," in *Models and Technologies for Intelligent Transportation Systems (MT-ITS), 2015 International Conference on*, Budapest, 2015.
11. D Dimov, "Crowdsensing: State of the Art and Privacy Aspects" [Online], September 2014. http://resources.infosecinstitute.com/crowdsensing-state-art-privacy-aspects/.
12. Google Play, "Dactyl—Fingerprint Camera" [Online]. https://play.google.com/store/apps/details?id=com.nyelito.dactyl&hl=en.
13. M A Alsheikh, D Niyato, D Leong, P Wang, and Z Han, "Privacy Management and Optimal Pricing in People-Centric Sensing," *IEEE Journal on Selected Areas in Communications*, vol. 35, no. 4, pp. 906–920, 2017.
14. N D Lane, S B Eisenman, M Musolesi, E Miluzzo, and A T Campbell, "Urban Sensing Systems: Opportunistic or Participatory?," in *Proceedings of the 9th Workshop on Mobile Computing Systems and Applications*, Napa, CA, 2008, pp. 11–16.
15. N Buluse et al., "Participatory Sensing in Commerce: Using Mobile Camera Phones to Track Market Price Dispersion," in *Proceedings of the International Workshop on Urban, Community, and Social Applications of Networked Sensing Systems (UrbanSense 2008)*, Raleigh, NC, 2008.
16. J A Burke et al., "Participatory Sensing," Center for Embedded Network Sensing, University of California, Los Angeles, 2006.
17. C Gao, F Kong, and J Tan, "Healthaware: Tackling Obesity with Health Aware Smart Phone Systems," in *Robotics and Biomimetics (ROBIO), 2009 IEEE International Conference on*, 2009.

18. P Jarvinen, T H Jarvinen, L Lahteenmaki, and C Sodergard, "HyperFit: Hybrid Media in Personal Nutrition and Exercise Management," in *Proceedings of the 2008 Second International Conference on Pervasive Computing Technologies for Healthcare*, Tampere, Finland, 2008.

19. A Sashima, Y Inoue, T Ikeda, T Yamashita, and K Kurumatani, "CONSORTS-S: A Mobile Sensing Platform for Context-Aware Services," in *Intelligent Sensors, Sensor Networks and Information Processing 2008 (ISSNIP 2008), International Conference on*, Sydney, Australia, 2008.

20. T Denning et al., "BALANCE: Towards a Usable Pervasive Wellness Application with Accurate Activity Inference," in *Proceedings of the 10th Workshop on Mobile Computing Systems and Applications*, New York, 2009.

21. S Consolvo et al., "Activity Sensing in the Wild: A Field Trial of Ubifit Garden," in *Proceedings of the SIGCHI Conference on Human Factors in Computing Systems*, New York, 2008, pp. 1797–1806.

22. X Chen, C T Ho, E T Lim, and T Z Kyaw, "Cellular Phone Based Online ECG Processing for Ambulatory and Continuous Detection," in *Computers in Cardiology*, IEEE, Durham, NC, 2007.

23. S Reddy et al., "Image Browsing, Processing, and Clustering for Participatory Sensing: Lessons from a DietSense Prototype," in *EmNets '07 Proceedings of the 4th Workshop on Embedded Networked Sensors*, Cork, 2007, pp. 13–17.

24. X Bao and R R Choudhury, "Movi: Mobile Phone Based Video Highlights via Collaborative Sensing," in *Proceedings of the 8th International Conference on Mobile Systems, Applications, and Services*, San Francisco, 2010, pp. 357–370.

25. E Miluzzo, N D Lane, S B Eisenman, and A T Campbell, "CenceMe—Injecting Sensing Presence into Social Networking Applications," in *European Conference on Smart Sensing and Context*, Berlin, 2007, pp. 1–28.

26. R De Oliveira and N Oliver, "TripleBeat: Enhancing Exercise Performance with Persuasion," in *Proceedings of the 10th International Conference on Human Computer Interaction with Mobile Devices and Services*, New York, 2008, pp. 255–264.

27. Y F Dong et al., "Petrolwatch: Using Mobile Phones for Sharing Petrol Prices," 2009, ResearchGate.

28. P Mohan, V N Padmanabhan, and R Ramjee, "Trafficsense: Rich Monitoring of Road and Traffic Conditions us ing Mobile Smartphones," TechReport, MSR-TR-2008-59, 2008, available at https://www.microsoft.com/en-us/research/publication/trafficsense-rich-monitoring-of-road-and-traffic-conditions-using-mobile-smartphones/.

29. T Das, P Mohan, V N Padmanabhan, R Ramjee, and A Sharma, "PRISM: Platform for Remote Sensing Using Smartphones," in *Proceedings of the 8th International Conference on Mobile Systems, Applications, and Services*, San Francisco, 2010, pp. 63–76.

30. N Maisonneuve, M Stevens, M E Niessen, and L Steels, "NoiseTube: Measuring and Mapping Noise Pollution with Mobile Phones," in *Information Technologies in Environmental Engineering*, Environmental Science and Engineering, Berlin, 2009, pp. 215–228.

31. A Thiagarajan et al., "VTrack: Accurate, Energy-Aware Road Traffic Delay Estimation Using Mobile Phones," in *Proceedings of the 7th ACM Conference on Embedded Networked Sensor Systems*, Berkeley, CA, 2009, pp. 85–98.

32. M Mun et al., "PEIR, the Personal Environmental Impact Report, as a Platform for Participatory Sensing Systems Research," in *Proceedings of the 7th International Conference on Mobile Systems*, Kraków, 2009, pp. 55-68.

33. K K Rachuri et al., "EmotionSense: A Mobile Phones Based Adaptive Platform for Experimental Social Psychology Research," in *Proceedings of the 12th ACM International Conference on Ubiquitous Computing*, Copenhagen, 2010, pp. 281–290.

34. N Oliver and F Flores-Mangas, "Healthgear: Automatic Sleep Apnea Detection and Monitoring with a Mobile Phone," *Journal of Communications*, vol. 2, no. 2, pp. 1–9, 2007.

35. Z Jin, J Oresko, S Huang, and A C Cheng, "HeartToGo: A Personalized Medicine Technology for Cardiovascular Disease Prevention and Detection," in *Life Science Systems and Applications Workshop 2009*, IEEE/NIH, Bethesda, MD, 2009, pp. 80–83.

36. J Dai, X Bai, Z Yang, Z Shen, and D Xuan, "PerFallD: A Pervasive Fall Detection System Using Mobile Phones," in *Pervasive Computing and Communications Workshops (PERCOM Workshops), 2010 8th IEEE International Conference on*, Mannheim, 2010, pp. 292–297.

37. F Sposaro and G Tyson, "iFall: An Android Application for Fall Monitoring and Response," in *Engineering in Medicine and Biology Society 2009 (EMBC 2009), Annual International Conference of the IEEE*, Minneapolis, 2009, pp. 6119–6122.

38. A Beach et al., "Whozthat? Evolving an Ecosystem for Context-Aware Mobile Social Networks," *IEEE Network*, vol. 22, no. 4, 2008, pp. 50–55.

39. M Klepal et al., "OLS: Opportunistic Localization System for Smart Phones Devices," in *Proceedings of the 1st ACM Workshop on Networking, Systems, and Applications for Mobile Handhelds*, Barcelona, 2009, pp. 79–80.

40. P Mohan, V N Padmanabhan, and R Ramjee, "Nericell: Rich Monitoring of Road and Traffic Conditions Using Mobile Smartphones," in *Proceedings of the 6th ACM Conference on Embedded Network Sensor Systems*, Raleigh, NC, 2008, pp. 323–336.

41. P Dutta et al., "Common Sense: Participatory Urban Sensing Using a Network of Handheld Air Quality Monitors," in *Proceedings of the 7th ACM Conference on Embedded Networked Sensor Systems*, Berkeley, CA, 2009, pp. 349–350.

42. IBM, "Creekwatch" [Online]. http://www.research.ibm.com/social/projects_creekwatch.shtml.

43. S B Eisenman et al., "BikeNet: A Mobile Sensing System for Cyclist Experience Mapping," *ACM Transactions on Sensor Networks (TOSN)*, vol. 6, no. 1, 2009.

44. J Yick, B Mukherjee, and D Ghosal, "Wireless Sensor Network Survey," *Computer Networks*, vol. 52, no. 12, pp. 2292–2330, 2008.

45. A Abduvaliyev, A-S K Pathan, J Zhou, R Roman, and W-C Wong, "On the Vital Areas of Intrusion Detection Systems in Wireless Sensor Networks," *IEEE Communications Surveys & Tutorials*, vol. 15, no. 3, pp. 1223–1237, 2016.

46. A-S K Pathan, *Security of Self-Organizing Networks: MANET, WSN, WMN, VANET*, CRC Press, Boca Raton, FL, 2016.

47. A-S K Pathan and C S Hong, "SERP: Secure Energy-Efficient Routing Protocol for Densely Deployed Wireless Sensor Networks," *Annals of Telecommunications*, vol. 63, no. 9–10, pp. 529–541, 2008.

48. X Shi et al., "Empirical Investigation on Safety Constraints of Merging Pedestrian Crowd through Macroscopic and Microscopic Analysis," *Accident Analysis & Prevention*, vol. 65, pp. 405–416, 2016.

49. J Noureen and M Asif, "Crowdsensing: Socio-Technical Challenges and Opportunities," *International Journal of Advanced Computer Science and Applications*, vol. 3, no. 9, pp. 363–369, 2017.

50. S Amin et al., "Mobile Century Using GPS Mobile Phones as Traffic Sensors: A Field Experiment" [Online]. http://citeseerx.ist.psu.edu/viewdoc/summary?doi=10.1.1.152.8548 (accessed 11 March 2018).

51. S Khan, A-S K Pathan, and N A Alrajeh. *Wireless Sensor Networks: Current Status and Future Trends*. Auerbach Publications, CRC Press, Taylor & Francis Group, Boca Raton, FL, 2012.

52. G Han, L Shu, A-S K Pathan, J Rodrigues, and A Mellouk, "Wireless Sensor Networks based on Environmental Energy Harvesting," Special Issue of International Journal of Distributed Sensor Networks, DOI: 10.1155/2013/816063, Volume 2013, Article ID 816063, Hindawi Publishing Corporation, 2013.

53. A-K A Abdalla and A-S K Pathan, "On Protecting Data Storage in Mobile Cloud Computing Paradigm," IETE Technical Review, Volume 31, Issue 1, 2014, pp. 82–91.

54. A-S K Pathan. Securing Cyber-Physical Systems. CRC Press, Taylor & Francis Group, Boca Raton, FL, 2015.

55. Suhas Mathur et al., "Parknet: drive-by sensing of road-side parking statistics," in *Proceedings of the 8th International Conference on Mobile Systems, Applications, and Services*, San Francisco, CA, 2010, pp. 123–136.

Chapter 2

Human Behavior in Crowd-Assisted Computing and Networking

Carlos Bermejo, Dimitris Chatzopoulos, Pan Hui, and Gunnar Karlsson

Contents

2.1 Crowdsensing

The capabilities of modern smart devices (SDs) provide new possibilities for more refined and precise studies in various areas due to their ability to contribute to crowd computing and networking procedures. These mobile devices have the potential to sense their surroundings with time, spatial granularity, and high accuracy in negligible time. Mobile SDs can contribute to the analysis of human behavior and natural phenomena such as pollution, CO_2 levels, traffic, and collective mobility patterns. Mobile crowdsensing is a paradigm that utilizes the ubiquitousness of the mobile users who are carrying smartphones and can collect and process data. Crowdsensing service providers request sensing tasks of mobile users who deliver these tasks in order to get paid. The participation cost of each mobile user is private information and depends on several factors. As a consequence, mobile users are motivated to misreport their actual cost in order to obtain higher payment, and hence incentives are needed for truth reporting. Crowdsensing tasks can be categorized based on characteristics inherent to the tasks or the participants. Two usual dimensions are event-based versus continuous, and spatial versus nonspatial. These dimensions are independent of each other, and any combination is possible. In this work, we focus on event-based spatial crowdsensing tasks. Such tasks are associated with geographic locations and whenever a mobile user receives one, she needs to perform the task in the defined location [1, 2]. SDs also have wireless capabilities that allow them to transfer the collected data to each other or to upload them to cloud servers. Figure 2.1 visualizes an example of a crowdsensing application where mobile devices can be interconnected and exchange collected data, which can be processed either locally, autonomously, or collaboratively, or uploaded to a server that handles the crowdsensing application.

A set of cooperation-enforcing mechanisms can be utilized depending on the requirements and the type of application, while the users of the crowdsensing application can impose their requests on the server and receive the results.

For instance, the data exchanged between mobile phones can inform users how long the queue is at a particular bus stop. Also, mobile crowdsensing mechanisms can monitor environmental conditions in a particular area, such as the humidity, pollen, and pollution, to provide meaningful information to a cloud service that detects potential health risks. However, due to the inherent human selfish behavior, the benefits of crowdsensing applications may not be clear to the participants in the field, and the requested tasks may not be executed due to low resource availability or trust issues. Given that any mobile user is self-interested, and he ideally uses others' resources/sensors without sharing any of his, cooperation-enforcing mechanisms have to be proposed and studies to improve users' altruism in crowd-assisted environments need to be conducted.

In this chapter, we initially present, via our experimental results, that crowdsensing users have a particular behavior regarding their location and confidence in the mobile crowdsensing application. We focus on task overhead performance [3]

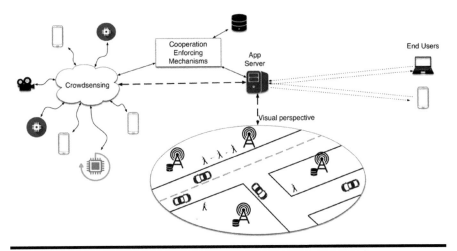

Figure 2.1   Devices are dynamically connected and are collecting and exchanging data. The crowdsensing application server utilizes the collected information and responds to queries. In order for the crowdsensing application to be functional, the mobile users have to be incentivized to dedicate parts of their resources.

and user altruistic behavior [4], respectively, to provide a better understanding of users' behavior in these mobile crowd–based scenarios. These insights can lead to non-monetary incentive techniques, such as better reliability and privacy settings on the server application that may concern users and degrade performance of the whole sensing ecosystem, or establish or predict the location of a particular user in order to request a sensing task and improve the acceptance ratio [3]. As we show in Section 2.2, mobile crowdsensing applications that consider user location and the trustworthiness of the service can boost their performance.

2.1.1 User Behavior

In mobile environments, when deciding whether to accept a task, users have to consider limited resources, such as remaining battery level, strength of social ties with other users, message content, and physical location. In this chapter, these individual decisions according to the previously mentioned parameters can impact the performance of crowd-based applications. Users' behavior can considerably impact crowd-assisted applications. How users react to task requests from other users based on their availability of resources and their social ties (i.e., reputation) affects the overall mobile crowdsensing performance. Resource availability is of high importance for mobile users, as they might not be able to use their devices for their needs, and hence crowdsensing tasks should not be assigned to mobile users when they do not have spare resources. For example, if a mobile user is at work, where she can charge her phone at any time and resources stay idle, a crowdsensing application can send a task to this user, as there will be available resources and the user will not reject the task. On the other hand, another user may be commuting and the availability of resources will be limited since she might not be able to charge her phone and may also be using her device at the same time; in this case, the probability of her rejecting the crowdsensing task is higher. As we will show in Section 2.2, users' behavior affects the overall performance of the acceptance probability of a task. The insights provided by this behavioral study will improve future crowd-based mobile applications and their respective incentives and routing mechanisms.

2.1.2 Altruism

The most important factor contributing to the success of data transmission applications running on crowd-assisted ecosystems is the goodwill of network users to compute other users' tasks (e.g., forward data packets to fellow users). For the last two decades, human altruism has been the target of several research fields, ranging from psychology to economics [5–8], most of which rely on some well-designed game, such as the prisoner's dilemma. However, the psychology of a user in an insulated lab environment is different from the normal state when we carry out daily activities. It is favorable to study the altruism issue in the context of a new technological environment. Altruism in these crowd scenarios has been studied in previous works [9–12], mainly from a routing perspective, focusing on the network performance under the assumption of different altruism patterns. Instead, in our work we intend to gain an understanding of human altruism from a practical point of view by taking into account factors like battery level and studying how these factors influence the altruism level in crowd-assisted networks. In mobile environments, when conducting altruistic behaviors, users have to consider limited resources, such as the remaining battery level, strength of social ties with other users, message content, and physical location. In this note, we focus on the study of the impact of battery level and social ties, as they play the most important role; other factors, like location and gender, are also considered.

2.1.3 Cooperation Mechanisms

Another important factor that affects the performance in crowdsensing applications is the trust between the mobile users and the crowdsensing task generators. Similar to human interactions, people behave in a more confident and relaxing way when they know the other person; this situation can be extrapolated to mobile crowdsensing scenarios, where the crowdsensing providers that are known by a user will have a higher probability of acceptance of their sensing task, and therefore improve their system performance. Furthermore, one of the features of mobile crowdsensing applications is the possibility of sensing and collecting data between peers (other mobile devices) without accessing the Internet; this provides aggregated data of the surroundings to all participants.

Cooperation-enforcing mechanisms can be categorized in many ways. The first way is based on their core component, which is the units that are employed to stimulate users' cooperation. These are *credits* and *reputation.*

Credit-based systems work as currencies. The key idea is that users providing a service should be remunerated, while nodes receiving a service should be charged. On the other hand, reputation-based systems are based on the trust that has built between the mobile users and their contribution to the community. Reputation-based schemes discourage misbehavior by estimating users' reputation and punishing the ones with bad behavior. The main difference between the reputation-based schemes and the credit-based schemes is that in the former each user can give a different trust score to a user, while in the latter all the users should know how many credits a user has. Any personal or collective reward can be of two types: monetary and non-monetary. Monetary rewards are easier for the users to understand and to examine whether it is beneficial for them to share their resources, but it is difficult from the crowdsensing mechanism to decide the monetary rewards. On the other hand, non-monetary rewards are much simpler for the crowdsensing mechanism to handle, but each user can give a different evaluation of such rewards. Figure 2.2 shows a categorization of different techniques that can be implemented. The baseline scenario in mobile crowdsensing applications is the one that does not incorporate any cooperation-enforcing mechanisms and the mobile users share their resources in an altruistic manner.

2.2 Users' Behavior

The main goal of our experiment is to depict the importance of the mobile users' state, in terms of location and resource availability, as well as their relationship with the other mobile users and the crowdsensing application provider. We designed

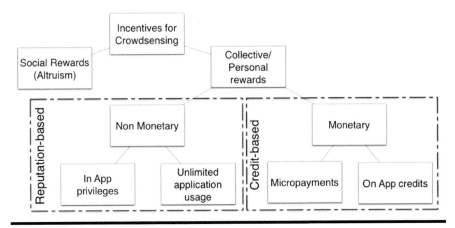

Figure 2.2   Categorization of cooperation-enforcing mechanisms.

a set of experiments to analyze these factors and examine whether it is feasible for a crowdsensing application provider to employ non-monetary incentive schemes to motivate mobile users to participate. Every crowdsensing task will consume users' limited resources (i.e., battery, processing power, memory, and Bluetooth interface), and this fact will affect the crowdsensing users' behavior. An example of a crowdsensing task can be the discovery of nearby devices via Bluetooth to measure the number of users in a particular area and use their accelerometer to detect possible traffic jams. Besides, we focus part of our analysis in altruism, as this particular behavior can affect performance, and also incentive mechanisms. Altruism can be seen in crowd-based environments as the best users' behavior as they accept any incoming task, and rewards can be nonexistent. This particular individual behavior can lead to big performance improvements if many users follow this pattern and, in the best-case scenario, provide the best scenario for crowd-based applications. This can lead to crowd-based solution adoption on mobile environments, where resources are scarce or limited.

We believe that the insights from this experiment can lead future work to provide better routing protocols and incentive schemes. Furthermore, prediction models for user location, resource availability, and users' behavior can open a new wave of innovative crowd-based applications, such as crowdsensing.

2.2.1 Testbed and Crowdsensing Requests

We developed an Android application that emulates a mobile crowdsensing application in order to perform our experiments. This application tracks users' locations when they reply to a sensing task request, and it simulates the trust of the ecosystem using the social context offered by Facebook (Facebook application programming interface [API]), which provides to the crowdsensing application a social network layer. Facebook offers the social ties (friendships) that play the same role as a trusted server or peer—user relation in a real crowdsensing scenario, whose impact will be equal in our analysis.

The application creates requests on users' devices to process or sense small tasks from other peers (Facebook social tie) with a corresponding battery consumption (0%–2%). The participants are requested to accept 20 tasks in a period that can vary from hours (request time span is 30–60 minutes) to days, depending on user task response rate. The request will show the current battery level, the size of the sensing task, the estimated battery level at the end of the task execution (battery consumed), and the social relationship with the sensing task owner (either a Facebook friend or an unknown peer). Each request shows the real devices' battery level; the battery cost (0%–2%) is a virtual measurement that does not imply an actual battery drain on users' devices. The visualization of the battery cost is a sufficient parameter to measure participants' response behavior; the real implementation of a background process that decreases battery is part of our future work. Figure 2.3 presents the design of the implemented testbed. The experimenters were recruited using Amazon Mechanical Turk (AMT), a crowdsourcing platform where they can

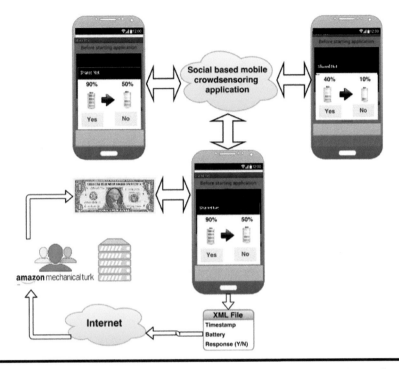

Figure 2.3 Implemented testbed for experimentation on cooperation-enforcing mechanisms on mobile crowdsensing.

download the Android application to their devices and receive a reward of $1 after they have finished the experiment. The demography of our experiment is shown in Figure 2.4.

2.2.1.1 Resource Availability

We initially examine the case regarding the resource availability when a crowdsensing task is executed and the location of the users' mobile devices. To track the users' locations, we collect the devices' GPS in the moment the users reply to the sensing task request; we have analyzed the location patterns of users during the experiment (i.e., user location does not change and corresponds to a particular house or building). The resource availability is related to the users' location, fixed or common locations such as work or office, home, or coffee shops that may offer to the users' battery-charging possibilities, and hence they affect their behavior and the mobile crowdsensing performance. We classify the availability of the resources into three groups:

■ **High:** Home, office, or any place that the user may stay for more extended periods and offer the possibility of charging the mobile device

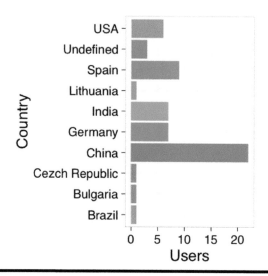

Figure 2.4   Demography of the users that participated in our experiment.

- **Restricted:** Coffee shops, gym, libraries, and places that offer battery-charging options, but that is not always the case
- **Limited:** Streets (walking), commuting situations, and scenarios where the user is moving or there is no charging infrastructure available

Figure 2.5 shows the distributions of users' resource availability for each sensing task request. The restricted availability (light blue), due to the probability of scenario chances, is lower in comparison with the other groups. We decide not to include any insights regarding this group as the collected data is not comparable to that of the other groups (highly and limited). We can see users' movement patterns, as limited (e.g., commuting) and high (e.g., home or office) contain most of our sample data. In most crowd-based scenarios, these extreme cases regarding resource availability play key roles for the success of mobile crowd–based approaches (e.g., crowdsensing).

We illustrate the users' location at the time of the generation of each crowd-sensing task request in Figure 2.6. We observe the users' movements along the experiment time span, and the different resource availability scenarios, and we illustrate how the users' movement behavior can provide the fundamentals to predict resource availability in the near future before a crowdsensing task request. We also observe that some users do not change their location (resource availability is stable) during the experiment. This can be due to fast response to each task request along the experiment duration. Crowdsensing applications in these cases can be limited regarding resource availability, as most of their scenarios require a particular location to perform their task, such as a bus stop or a specific area of a city. Therefore, we analyze their crowdsensing performance in both situations.

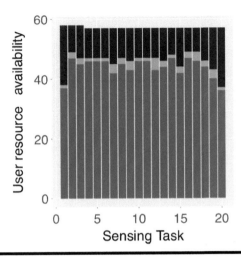

Figure 2.5 Highly (baseline blue), restricted (lightest blue), and limited (darkest blue) resource availability according to sensing task requests.

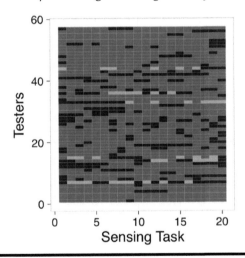

Figure 2.6 Heatmap of resource availability for each user at the moment of the crowdsensing task generation. The darkest blue corresponds to the highly group, base blue to the restricted, and lightest blue to the limited.

The next figures focus on analysis regarding resource availability: high and limited. Figure 2.7 shows the scenario where a user executes the sensing task in a location where she has a chance to charge her device. It illustrates the probability of acceptance, which is the probability of the user executing the crowdsensing task. We can see that on average this probability is around 42%. Moreover, the probability of acceptance does not change with time (number of sensing task requests). In Figure 2.8 we observe the opposite case, where users do not have charging

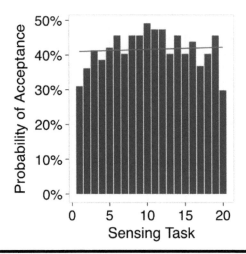

Figure 2.7 **Probability of a mobile user accepting a crowdsensing task in the case of having available resources.**

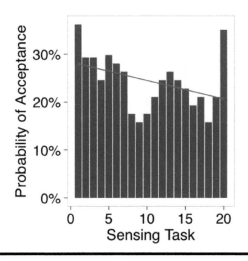

Figure 2.8 **Probability of a mobile user accepting a crowdsensing task in the case of not having available resources.**

possibilities. The probability of acceptance is lower compared with the case of existing available resources (highly); its average is around 25%, and the probability of executing a task decreases with the number of sensing task requests. Due to the nature of acceptance applications, the battery consumption will be progressive, and hence will decrease device battery life with the number of sensing task requests. Moreover, increasing the number of requests affects the performance on the probability of acceptance, and therefore the mobile crowdsensing application.

The more a user accepts sensing task requests, the more correlated is the availability of resources for this particular user. Therefore, applications need to consider the number of requests a user is willing to accept according to their resources. This can lead to better routing mechanisms as it will consider the number of accepted requests and the availability of resources of the following user. For example, if a user accepted several requests and her resource availability is low, the routing protocol will forward a task to another node who has either more resources, less accepted sensing tasks, or both. For the previously mentioned example of a crowdsensing task in a particular city area, the system needs to consider the number of task (sensing) requests per user and try a wider spectrum of users instead of multiple requests to the same users. This approach can raise sensing accuracy with more refined data and avoid any user overhead due to the low probability of acceptance of a task.

The study of a user's location (resource availability) leads to non-monetary incentive techniques that can improve the performance of the whole crowdsensing ecosystem. Such applications require analysis of a user's location and previous sensing task requests in order to offer the sensing task when the user is, for example, in the highly group, as the probability of acceptance is higher. Furthermore, the crowdsensing application can schedule the sensing task in situations where the app knows that the user is in a particular location where she can charge her mobile device (user mobility patterns and prediction). As we will mention in other aspects during this chapter, prediction can lead to big performance improvements if the prediction model considers factors such as location, resource availability, and user's previous behavior.

2.2.1.2 User Interrelationship

In this section, we aim to analyze the importance of social ties (trust) among crowdsensing application participants (servers or peers) to boost the system performance. We emulate our social layer using Facebook API, which offers the required social relationships between a user and crowdsensing task creator or owner (peer or server): our mobile social-based crowdsensing application. The developed Android application emulates the task request, and it will show the battery consumption and social tie with the task owner or creator.

For the social ties (user interrelationship), we consider the following two cases:

- **Case 1:** The user or participant knows the task creator or owner.
- **Case 2:** The user or participant does not know the task creator.

The emulation of the server or peer trust using Facebook offers a feasible solution for our experiment and a reliable simulation of a real social mobile crowdsensing application. As we present in this section, the application's performance increases with the trust (social relationship) between participants.

Furthermore, in this experiment scenario we also focus our interest on the *task overhead*: the number of resources (i.e., battery, processing, and memory) that a particular task will consume in a user's device. As the user location cannot be fixed by the crowdsensing application (only schedule the task request), the overhead can be decreased with non-monetary incentive techniques, such as server or peer trust, which can be controlled by the mobile crowdsensing application.

Figure 2.9 shows the probability of acceptance regarding the task overhead. We observe the linear relationship between both parameters, and how the task overhead affects the users' behavior and the crowdsensing task execution. The probability of acceptance is approximate 55%. The most important aspect is the difference when the task overhead is near 0.5% (battery in our experiment scenario), which has a 62% probability of acceptance, as opposed to a 2% task overhead, which has a 47% probability of acceptance. This points out the importance of the task overhead in a mobile crowdsensing ecosystem, and how applications have to target low resources consumption (battery, memory, and CPU) in order to achieve better performance. One solution to reduce the impact of task overhead in the performance of mobile crowdsensing applications is the addition of the commented social layer or trusted servers or peers to the system. This can be seen as a non-monetary incentive technique (cooperation-enforcing mechanisms) that can improve the performance on the probability of crowdsensing task acceptance. In Figure 2.10, we illustrate the probability of acceptance of a particular crowdsensing task with its respective overhead in scenarios where the participants' trust (case 1) the task owners or creators. If we compare it with Figure 2.11 (case 2), we can observe the difference in the probability of acceptance of a crowdsensing task. The addition of a social layer in our experiment, and therefore trusted servers

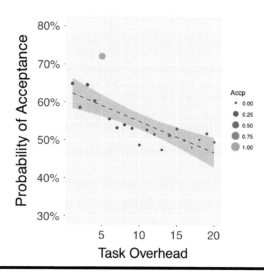

Figure 2.9   All requests.

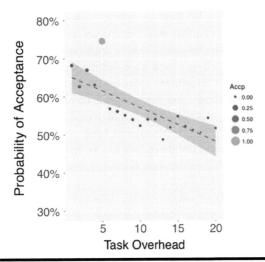

Figure 2.10   Requests between friends.

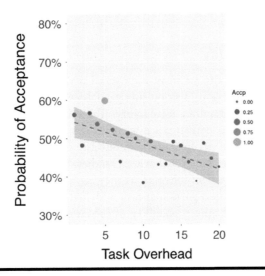

Figure 2.11   Requests between unknowns.

or peers in a real application, can boost the performance in scenarios where the task overhead can be the main constraint in users' or participants' willingness to execute a task.

In this section, we demonstrated how the resource availability and trust in servers or peers, both feasible non-monetary incentive techniques to implement, can improve the cooperation, and hence performance, of mobile crowdsensing applications. Moreover, we described some users' patterns regarding their location (resource availability) and behavior when the server or a peer is not a trusted source (crowdsensing task creator). We have to highlight the constraint that task overhead can become in mobile scenarios, where resources are limited. All these insights can be the foundation of better crowdsensing applications, which can employ non-monetary incentive techniques, such as enforcing mechanisms (trusted servers), and the collection of users' locations. These results prove that social mobile crowd-sensing applications can boost the performance of baseline mobile crowdsensing applications. However, it is not always possible to find mobile users that are socially close to the crowdsensing task initiator and have available resources to help. In the next section, we discuss how incentives can be utilized to attract more mobile users and how the proper selection of mobile users can improve the efficiency of the crowdsensing mechanism.

However, in crowd-based environments there can always be agents who accept tasks from others without any retribution expectation (i.e., altruistic behavior). Therefore, we need to study this particular behavior for crowdsensing environments as it can lead to performance improvements with little incentive mechanisms. Altruistic behaviors can offer best scenarios for crowd-assisted computing and networking ecosystems, as they will not require complicated incentive schemas and users will accomplish tasks without a *reward goal*.

2.2.1.3 Altruism Study

Users' behavior studies can improve the overall performance of current crowd-based systems. The previous section analyzed how users behave according to the task overhead and resource availability. There is still a particular behavior that can improve crowd performance with no cost for incentive mechanisms. Altruism causes users to accept tasks without any reward goal. Altruistic users play an important role in crowd behavioral studies. These users can also modify current routing or assignment schemas, as they are always willing to help for the cause (i.e., sense or execute a particular task with little or no reward as the target). In this section, we study the probability of acceptance according to one of the main constraints in crowd-assisted mobile environments: *battery level*. Battery is still one of the main concerns for mobile users, as it is one of the weakest features for mobile environments. Therefore, we simplify this altruism study with only one factor, besides inherit psychological ones, to evaluate users' willingness to accept a task.

Results shows that there is an altruistic–selfish phase transition when battery power is less than 40%. This phase transition can be used to improve future incentive mechanisms as its importance affects users' behavior and will negatively affect any incentive mechanism approach. Future work can include this particular battery point to avoid wasting networking and computing resources in their incentive schemas, as the probability of acceptance (i.e., users' willingness) will be so low that the incentives will need to increase to compensate, even making them *not worth the try*.

Furthermore, these results help us to better understand human behavior in crowd-assisted environments and will help future work find innovative ways to improve and manage pervasive computing systems that require the active role of users.

In Figure 2.12, we give the relationship between averaged cumulative acceptance probability, $p^f \leq (l_{BL})$, and battery level, l_{BL}. As the battery level is discrete and data is sparse, so the averaged cumulative acceptance probability would be a good quantity to characterize the connection between forwarding probability and battery level. As shown in the picture, the average acceptance probability for the whole range from 26% to 90% is nearly 70%; however, when the battery level is lower than 45%, users are much less altruistic. The 40%–50% battery level can be viewed as a critical region on which a user's altruism changes radically below and above this region—an altruistic–selfish phase transition lies inside this region. This critical section is very important for future works on incentive mechanisms, as it can limit the incentive approaches for situations where users do not have battery-level resources above the critical region (40%–50%).

To study a user's behavior under different battery statuses, charging or discharging, we split the data based on the device battery status, getting the $p^f \leq (l_{BL}) - l_{BL}$ relation for each status, which is given in Figure 2.13. We can see that in the battery range above 70%, there is not much difference between these two statuses, while below 70% until 50%, users using a charging SD start to show higher altruism.

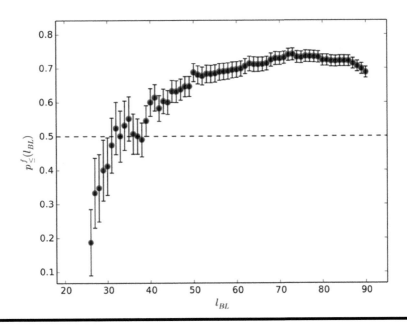

Figure 2.12 Averaged cumulative acceptance probability with battery level. $p^f \leq (l_{BL})$ means the average acceptance probability over battery levels that are $\leq l_{BL}$.

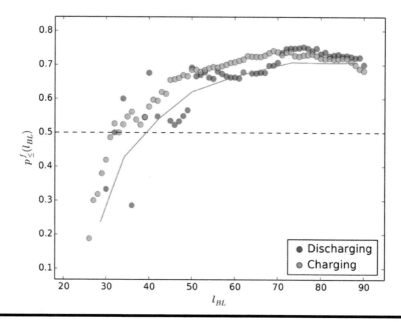

Figure 2.13 Averaged cumulative acceptance probability with battery level for different battery states.

Unfortunately, when the battery power is less than 50%, people are likely to charge their devices, which makes the data of charging status in the below 50% range so sparse that we cannot make a fair comparison with data of the discharging status, but we believe the difference of altruism value between these two states will keep growing as battery power goes down.

The differences in acceptance probability after the battery level goes below the critical region can benefit future incentive schemes. For example, an incentive approach that takes into account the mobile device charging status can vary according to it and achieve more redefined schemes. Moreover, predicting the charging status of mobile devices on a crowd-based application can provide better performances, as the incentive mechanisms can evaluate *best-scheme approaches* based on not only current status but also future device parameters (e.g., battery will be plugged in for the next 5 hours).

Figure 2.14 illustrates the different altruism distributions with respect to different social ties cases. Some points have zero error bar because there is only one record in that specific battery level and social ties case. For the whole range of battery level, we considered social case 2, which corresponds to an unknown task creator and has the lowest acceptance probability, while social case 1, which corresponds to a known task creator, has the highest acceptance probability. It is easy to understand as we are more willing to forward messages that come from someone we know. Moreover, the mentioned critical region of 40%–50% affects both cases regarding acceptance

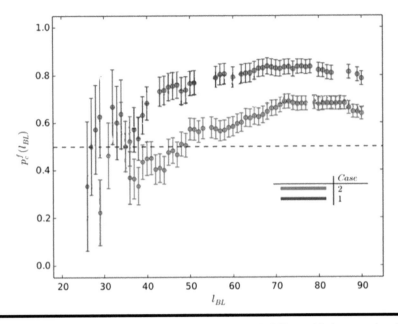

Figure 2.14 Averaged cumulative acceptance probability with battery level for different social ties cases.

probability of a task. This case is easier to depict in an unknown server. Therefore, reputation and incentive schemas are necessary for crowd-based scenarios as users' previous behavior affects not only their goals (i.e., the probability of acceptance task success in other devices) but also the overall crowd scenario performance.

In this section, we provide a detailed analysis of users' behavior and how experiment results can give insights for future incentive mechanism schemas for mobile crowd–based applications. We also found critical regions where incentive schemas either will have to increase their reward considerably or will not work due to users' acceptance probability. Moreover, we found that social ties affect users' responses and willingness to compute, sense, or execute another user's task. We also studied users' altruism as one key factor that can lead to non-highly-incentive-dependent crowd-based scenarios, where users just accept other users' tasks without any reward expectation. Therefore, reputation and incentive schemes are necessary to achieve better performance results in these crowd-based mobile scenarios. In the following section, we focus on adaptation of social ties to social-based mobile crowdsensing applications.

Future routing protocols and systems for crowdsensing applications can consider these results in their approaches, as they can improve not only users' adoption of crowd-based systems but also data in the case of crowdsensing applications or performance in offloading systems (i.e., computing offloading to nearby devices).

2.3 Adaptation to Social-Based Mobile Crowdsensing Application

Although social ties between mobile users can affect the performance of a mobile crowdsensing application, if the initiator of the task is a cloud service, the mobile users should be rewarded in order to be cooperative. The social relationships can improve the performance of the application in the case where a mobile user has received a crowdsensing task and she offloads parts of it to her neighbors in order to be able to provide a more robust result to the requester.

The sustainability of cooperation enforcing in crowdsensing is aligned with the architectural design of the crowdsensing mechanism. The number of available devices that can sense a multitude of parameters of cyber physical space can produce huge amounts of data that are highly correlated and should not be uploaded to the cloud or exchanged between SDs unfiltered. The selection of which users to select for uploading their data is not trivial or static. However, the social relationships (trust) between the mobile users can help in the selection of the most suitable ones (the most popular).

The high variance on the possible crowdsensing applications makes reputation-based incentives not applicable for a generalized crowdsensing mechanism that is based on mobile devices. The core reasoning behind this is the fact that the service exchange is asymmetric, in the sense that users are sharing their resources in order to earn something and are not motivated to build a good reputation in order to utilize the service later. A second reason is the wide spectrum of the possible crowdsensing tasks. It is difficult to determine a representative valuation for a credit-based cooperation-enforcing mechanism for a generalized crowdsensing mechanism. A habitual approach can be based on auctions. Whenever there is a crowdsensing task, the application provider broadcasts the characteristics of the task he wants to execute, and the interested mobile devices bid for the task and propose their valuation. However, this requires the implementation of a component to the mobile devices that would be able to estimate the evaluation of the task and bid for it. The implementation of such a software component is not trivial and takes time to execute each time.

Related works, such as [13–16], assign a specific cost per action that does not depend on the user that performs the task. Any credit-based cooperation-enforcing mechanisms have to guarantee consistency, and the difficulty of this depends on whether they use any complementary infrastructure. One case is where the mechanism uses a remote service for bookkeeping. At the end of each interaction, only two participants need to connect to the remote server in order to update their credit score. In the case of absence of a remote service, the mechanism has to integrate a local storage service on each mobile device and guarantee consistency. However, this approach requires too many messages to be exchanged in order to inform every node in the crowdsensing application about any transaction. A decentralized cooperation-enforcing mechanism for a large-scale crowdsensing mechanism is not implementable without the use of Internet connectivity and a bitcoin-like proof-of-work scheme that may require many resources from the participants and constrain its implementation [17].

Furthermore, credit-based schemes for crowdsensing applications with a variety of tasks have to deal with two more issues that are not important in application-specific credit-based schemes. Given that different tasks may have different valuation and different users may have different needs, the credits may accumulate to a subset of users that have available computational resources to help others, but they do not frequently ask for help from others. Also, users' mobility pattern and configuration may not help them gain credits.

2.4 Conclusion

In this chapter, we discussed what kind of cooperation-enforcing mechanisms implemented for mobile ecosystems can be adapted in crowdsensing applications. Furthermore, we discussed the required complementary infrastructure by cooperation-enforcing mechanisms for both centralized and distributed approaches. These mechanisms can heavily affect the impact of mobile crowdsensing scenarios.

We justified our arguments via experimentation with real users. The insights we obtained from the experiments show how a collective reward (social relationships) can change users' psychological status and thus lead to largely different participation behavior. This suggests that a proper incentive mechanism should be introduced into mobile crowdsensing applications to improve their feasibility. Another important fact that has been presented is the existence of noncooperative behavior in cases where the crowdsensing task overhead increases. We have shown that participants are relatively more cooperative and have a higher probability of acceptance when their devices are in zones with high resource availability (device-charging possibilities). The combination of the cooperative mechanisms together with appropriate user incentive schemas will boost the performance and adoption of mobile crowdsensing applications. More sophisticated incentive techniques, which can be implemented on top of our concept of social mobile crowdsensing applications, are part of our future work. We found that the average acceptance probability is 70%, which shows how a minor reward ($1 in this case) can change a user's psychological status and thus lead to largely different altruistic behavior. This suggests that a proper incentive mechanism should be introduced into crowd environments to improve crowd-assisted feasibility. An important fact that has been presented is the existence of an altruistic–selfish phase transition with regard to the remaining battery level, where a small variation of this parameter can completely change the user's behavior and his replies. A battery level lower than the critical region (40%–50% in this case) gives a much lower altruism value. This phase transition from altruistic to selfish behavior will impact future decision-routing and incentive schema mechanisms. We also show that users are relatively more altruistic when their devices are charging than discharging. How to keep a user's battery level above the critical region for as long as possible is an urgent issue for research.

The advantages brought by an efficient mobile crowdsensing environment may outweigh the cost spent on implementing new technologies, such as mobile charging, in order to maintain those key parameters above the critical region. Social ties in our research are found to play either a positive or a negative role in enhancing altruism, depending on to what extent the forwarders think it is useful to forward the message. So, the message quality and customization for users are very import factors in the implementation of crowd-assisted computing and networking scenarios. At the same time, we found that to make users prefer altruism rather than selfishness, the cost of an altruistic action should not be lower than 10% of the resource owned by the user. However, we need to accurately define the cost for a real user in our future work.

Future work needs to focus on more redefined incentive mechanisms, where current and estimated resource availability, such as future location and battery status (i.e., charging or discharging), need to be considered. Furthermore, reputation schemes need to analyze users' behavior according to knowledge about the task creator, and neighbors. Moreover, these reputation mechanisms need to promote altruism, as it provides the best case for the success of crowd-based applications. Routing protocols can also benefit this work, as our results demonstrate the probability of acceptance of a particular task or, in the case of offloading scenarios, the forwarding of a task among neighbors (i.e., which users should be the next hop to forward and execute a task).

To conclude, we have conducted a detailed experiment to study users' behavior in crowdsensing applications from an empirical point of view, and through a wide analysis of the results, we get the dependency of the probability of acceptance according to the task overhead parameter and the average altruism on parameters like remaining battery level, social ties, and battery status. This work can lead to future implementation of incentive mechanisms to improve not only users' altruistic behavior but also the overall performance in crowd-based scenarios.

Acknowledgments

This chapter is supported by the Sponsorship Scheme for Targeted Strategic Partnership (SSTSP) of the Hong Kong University of Science and Technology (HKUST).

References

1. R. K. Ganti, N. Pham, H. Ahmadi, S. Nangia, and T. F. Abdelzaher. GreenGPS: A participatory sensing fuel-efficient maps application. In *Proceedings of the 8th International Conference on Mobile Systems, Applications, and Services*, MobiSys '10, New York, 2010, pp. 151–164.

2. T. Yan, B. Hoh, D. Ganesan, K. Tracton, T. Iwuchukwu, and J.-S. Lee. Crowdpark: A crowdsourcing-based parking reservation system for mobile phones. Technical Report. University of Massachusetts, Amherst, 2011.

3. C. Bermejo, D. Chatzopoulos, and P. Hui. How sustainable is social based mobile crowdsensing? An experimental study. In *2016 IEEE 24th International Conference on Network Protocols (ICNP)*, Singapore, 2016, pp. 1–6.

4. C. Bermejo, R. Zheng, and P. Hui. An empirical study of human altruistic behaviors in opportunistic networks. In *Proceedings of the 7th International Workshop on Hot Topics in Planet-Scale Mobile Computing and Online Social Networking*, Heidelberg, Germany, 2015, pp. 43–48.

5. E. Fehr and U. Fischbacher. The nature of human altruism. *Nature*, 425(6960), pp. 785–791, 2003.

6. E. Fehr and S. Gächter. Altruistic punishment in humans. *Nature*, 415(6868), pp. 137–140, 2002.

7. S. Levitt and S. J. Dubner. *Super Freakonomics*. William Morrow: New York, 2009.

8. D. K. Levine. Modeling altruism and spitefulness in experiments. *Economic Dynamics*, 1, pp. 593–622, 1998.

9. P. Hui, K. Xu, V. O. K. Li, J. Crowcroft, V. Latora, and P. Lio. Selfishness, altruism and message spreading in mobile social networks. In *Proceedings of the IEEE INFOCOM Workshops*, Rio de Janeiro, Brazil, 2009, pp. 1–6.

10. Q. Li, S. Zhu, and G. Cao. Routing in socially selfish delay tolerant networks. In *IEEE INFOCOM*, San Diego, CA, 2010, pp. 857–865.

11. Y. Li, P. Hui, D. Jin, L. Su, and L. Zeng. Evaluating the impact of social selfishness on the epidemic routing in delay tolerant networks. *IEEE Communications Letters*, 14(11), pp. 1026–1028, 2010.

12. J. Su, J. Scott, P. Hui, J. Crowcroft, E. de Lara, C. Diot, A. Goel, M. Lim, and E. Upton. Haggle: Seamless networking for mobile applications. In *Proceedings of ACM UbiComp*, Innsbruck, Austria, 2007, pp. 391–408.

13. S. Zhong, J. Chen, and Y. R. Yang. Sprite: A simple, cheat-proof, credit-based system for mobile ad-hoc networks. In *INFOCOM, IEEE*, San Francisco, CA, 2003, vol. 3, pp. 1987–1997.

14. B. B. Chen and M. C. Chan. Mobicent: A credit-based incentive system for disruption tolerant network. In *INFOCOM, IEEE*, San Diego, CA, 2010, pp. 1–9.

15. H. Zhu, X. Lin, R. Lu, Y. Fan, and X. Shen. Smart: A secure multi-layer credit-based incentive scheme for delay-tolerant networks. *IEEE Transactions on Vehicular Technology*, 58(8), pp. 4628–4639, 2009.

16. J. Crowcroft, R. Gibbens, F. Kelly, and S. Östring. Modelling incentives for collaboration in mobile ad hoc networks. *Performance Evaluation*, 57(4), pp. 427–439, 2004. Available: http://dx.doi.org/10.1016/j.peva.2004.03.003.

17. S. Nakamoto. Bitcoin: A peer-to-peer electronic cash system. 2009. Available: http://www.bitcoin.org/bitcoin.pdf.

Chapter 3

IoT-Based Smart Surveillance: Role of Sensor Data Analytics and Mobile Crowd Sensing in Crowd Behavior Analysis

Sabu M. Thampi and Elizabeth B. Varghese

Contents

3.1 Introduction

The Internet of Things (IoT) is the interrelated networking of devices, objects, digital machines, people, or anything, to communicate, exchange data, and so forth, without human intervention. The objects can capture, communicate, store, access, and share data from the physical world. IoT can be used in a big range of domains, such as predictive maintenance, visualization, logistic tracking systems, home automation, public surveillance or telematics, and remote device configuration management. With an extensive rise in terrorism, coupled with challenging security conditions, research in security and surveillance has become indispensable. In the field of security and surveillance, IoT-based applications can be utilized to monitor movements remotely and send out alerts when they are identified.

Employing video cameras for monitoring and surveillance is a conventional practice in both government agencies and private firms around the globe. Most current surveillance systems necessitate a human operator to constantly monitor them. Their effectiveness and response are largely determined by the vigilance of the person monitoring the camera system. Furthermore, the number of cameras and the area under surveillance are limited by the personnel available. To overcome these limitations, traditional surveillance methods have been ousted with smart surveillance systems. In smart surveillance systems, the system becomes smart by adding additional intelligence to the connected devices so that they can make decisions on their own. The smart surveillance system can be used for both indoor and outdoor applications. With the advent of IoT, indoor surveillance systems such as home automation systems are very common today. Figure 3.1 shows the general classification of a smart surveillance system. In both indoor and outdoor applications, smart surveillance can be implemented with three main approaches: *visual based*, *sensor based*, and *multimodal*.

■ Visual approaches use visual sensors, cameras, and unmanned aerial vehicles (UAVs) for surveillance. The main characteristics include processing digital data and automatically modifying extrinsic and intrinsic parameters, such as focus, iris, pan, tilt, and zoom, for increasing the data quality [1].
■ Sensor-based surveillance uses different types of sensors other than visual sensors for monitoring. They include motion sensors, accelerometers, actuators, and wearable sensors. They can track details of human behavior, including human body information such as actions and gestures.

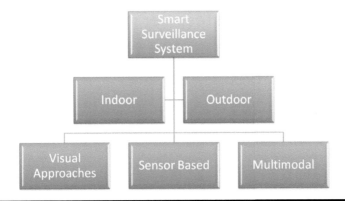

Figure 3.1   General classification of smart surveillance system.

■ Multimodal approaches use the integration of sensors and other modality data for surveillance. The data include mobile crowd sensing data, location details and phone call records, social network data, and data from radio frequency identification (RFID). Biometric details, such as fingerprints, DNA, and face patterns, can also be incorporated.

Nowadays, almost all urban areas are facing a wide range of threats, ranging from terrorism and civil unrest to kidnapping and murder. Moreover, because of the increase in the human population, crowd-related disasters are also common. To reduce the impact of such threats and disasters, it is critical for authorities to capture real-time information on what is happening in and around the city. This can be done by the adoption of information and communication technologies (ICT) and the IoT. This necessitates the deployment of a wide range of sensors and smart cameras in public places for real-time information capturing. Thus, the demand for IoT-based smart surveillance is increasing in areas ranging from crime prevention, public safety, and home security to industrial quality control and military intelligence gathering.

Monitoring and analyzing crowded scenes is one of the biggest challenges in a public surveillance system. Over the years, many crowd-related fatalities have occurred around the globe, in connection with music concerts, sports events, pilgrimages, and so on. Figure 3.2 shows the statistics of the major crowd disasters [2]. Many people lost their lives due to these major and minor accidents. The fact that the year 2017 experienced the most number of disasters is evident from the figure.

According to the report of the National Disaster Management Authority (NDMA) of the Government of India, a lack of effective surveillance for analyzing changes in crowd behavior is one of the major reasons for crowd-related disasters. Due to the increase in the human population and its diverse activities, crowded scenes are quite common today. Whether crowds gather to protest or just

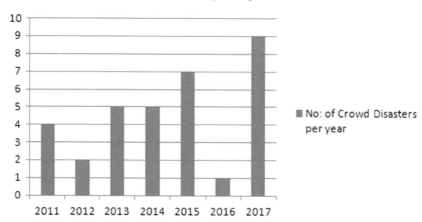

Figure 3.2   Statistics of crowd disasters [2].

for entertainment in a party, things may turn to be disastrous at any moment. Sometimes this may lead to fatal accidents. In order to avoid such disasters, crowd behavior analysis and proactive detection of suspicious human activities are absolutely necessary.

For analyzing the crowd behavior in a smart surveillance environment, one of the significant modalities of data is from mobile crowd sensing [3]. Today's smart phones are equipped with multiple sensors, and they can be employed as a personal sensing device. Mobile crowd sensing takes advantage of common people, to contribute data from their mobile devices. In a smart surveillance environment, the combination of traditional video and sensor data, along with mobile crowd sensing, can further enhance the possibility of detecting the anomalies in a crowd before a situation becomes worse.

This chapter is intended to provide an extensive review of the various analytics methods from video as well as sensor data for IoT-based smart surveillance. Section 3.2 discusses the different video analytics methods for analyzing crowd behavior and detecting abnormal activities. The different approaches for sensor data analytics are demonstrated in Section 3.3. Section 3.4 provides an outlook of various mobile crowd sensing approaches for crowd management. Finally, we conclude the chapter by presenting future work, including the essentiality of combining multimodal data.

3.2 Smart Surveillance from Video Data

One of the main applications of IoT is in the area of public surveillance. Public surveillance and security are essential for the government to protect citizens and organizations. Traditional closed-circuit television (CCTV) cameras are used only for capturing and storing video footage for later analysis. In contrast, smart cameras with visual sensors in an IoT-based surveillance use artificial intelligence (AI)–based programs to analyze the captured video and act accordingly. The analysis of video data by applying computer vision methods and creating real-time intelligence appropriate for the observed environment is called *video analytics*. In a public surveillance system, video analytics helps to detect unusual movements, breaking of traffic rules, parking in unauthorized areas, and so forth. For example, smart cameras will detect cars parked in "no parking" areas and send an alert to the driver through an attached speaker. If there is no response, they will inform officials to take action. The following sections discuss the surveillance methods in literature based on video analytics.

3.2.1 Video Analytics

Video analytics is a powerful tool that has the capability to convert unstructured video data into structured useful data that can be analyzed, searched, and managed to create a real-time intelligent response system. To increase the potential of video analytics solutions, computer vision algorithms are employed.

Video analytic methods for public surveillance are commonly used in intelligent traffic monitoring. The smart surveillance system proposed in [4] makes use of data from different types of sensors for license plate recognition, detecting suspicious behavior in parking lots, face detection and recognition, and badge reading for access control. This system uses IBM Middleware for Large-Scale Surveillance (MILS) and data management, which includes a set of web services. Fernández et al. [5] proposed an IoT surveillance platform, based on the usage of a large number of visual sensors. It facilitates data collection from the sensors and also provides alarm and video stream distribution toward the emergency teams. Here the visual sensors produce only XML output instead of video. The XML data are processed to create knowledge-based ontology for a semantic engine, for route detection of vehicles, and to analyze the abnormal trajectories.

In [6], a smart camera solution based on video analytics is used for an intelligent transportation system. The features include make and model recognition, license plate recognition, and color recognition of vehicles. The make and model are recognized using Speeded Up Robust Feature (SURF) descriptors, whereas the license plate is recognized using the Tesseract Optical Character Recognition (OCR) tool. The color recognition algorithm is based on dominant color analysis, performed in the lab space. A surveillance system using a distributed wireless network of smart cameras, modeled as agents, is proposed by Eigenraam and Rothkrantz [7]. This system is used for detecting illegal parking, speed limit crossing, and vehicles entering one-way streets and other scenarios of traffic rule violations. Since cameras are considered agents, a rule-based approach is used for the communication between agents and also to detect suspicious vehicles.

Many smart camera vendors are providing video analytics solutions for public surveillance. Some of the major smart camera vendors and features of their cameras are presented in Table 3.1. The features include intelligent video analysis, detection of accidents and emergency alert systems, traffic management through optimization of routes, line-crossing detection in restricted areas, loitering and idle object detection, automatic license plate recognition, trajectory tracking of vehicles, giving live feed to officers, data sharing via the cloud, detection of zero blind spots in surveillance networks, and face recognition [4, 8]. In order to make Bhubaneswar, India, a smart city, the smart cameras are used for loitering detection, detection of abandoned objects, alerting the police about traffic violations and abandoned vehicles, automatic number plate recognition, and email and mobile alerts [9]. In Austin, Texas, smart cameras are used for gunshot detection, license plate recognition, radioactive isotope detection, intelligent routing, and much more [10].

Another application area of public surveillance that utilizes video analytics is monitoring individuals and their activities, especially in crowded areas where there are chances for disasters and crime-related incidents. At events like concerts or public places such as pilgrimage spots or airports, where there is a large number of people, a pedestrian crowd monitoring and crowd disaster prevention system is necessary [11]. The increasing rate of crime, terrorism, and disasters in crowded places has motivated security organizations to develop robust smart surveillance systems. Although video analytics solutions are present in vehicle surveillance, motion detection, motion tracking, gunshot detection, face recognition, and so forth, very few of them analyze the behavior of people in a crowd, which is a complex task.

Table 3.1   Major Smart Camera Vendors and Their Solutions

Major Vendor	Solution
Video intelligence solution by IBM [4]	Detection of accidents and emergency alert systems, traffic management through optimization of routes, automatic license plate recognition, face recognition
Honeywell Technologies [9]	Loitering detection, detection of abandoned objects, alerting of police about traffic violations, automatic number plate recognition
Intel along with Hitachi [10]	Gunshot detection, license plate recognition, radioactive isotope detection, intelligent routing
Bosch [8]	Line-crossing detection in restricted areas, loitering and idle object detection, trajectory tracking of vehicles
iOmniscient Pvt Ltd. along with Cisco [12]	Suspicious human behaviors, like falling, fighting, loitering, and running
NEC Vision [13]	Sudden formation of crowd and crowd density heatmap

3.2.2 Crowd

A crowd can be defined as a group of individuals gathered in a single place at a particular time for a specific purpose [14]. It can be generally classified into two groups, based on the motion of the crowd.

- Stationary crowd: When the people are not moving from their place, this is referred to as a stationary crowd. They can be found as spectators or audiences at concerts, rallies, performances, and speeches. A stationary crowd is also formed when moving pedestrians stop walking and form groups. Figure 3.3 shows a stationary crowd attending a campaign.
- Dynamic crowd: A group of people on the move is referred to as a dynamic crowd. A dynamic crowd can be classified as structured or unstructured depending on their direction of motion. If the people in the crowd are moving in the same direction, the crowd is structured. Crowds appearing in marathons, rallies, and campaigns are structured. In an unstructured crowd, people move in all directions, such as in the case of shopping malls and airports. Structured and unstructured crowds are shown in Figure 3.4 [15].

Figure 3.3 Stationary crowd.

Figure 3.4 Structured and unstructured crowd [15].

A crowd can also be classified as active or passive based on the behavior of the people in the crowd. An audience who is not taking part in any action is considered passive. All others in the crowd are termed active. An active crowd can be divided into aggressive, escapist, acquisitive, and expressive. Figure 3.5 shows the characteristics of an active crowd. An aggressive crowd is violent and outwardly focused. Panicked people are called escapists, since they want to escape from their current situation. A large number of people fighting for limited resources are called an acquisitive crowd. Groups gathering for an active purpose, such as religious revivals and rock concerts, are expressive crowds. So, crowd behavior analysis is necessary to analyze and identify an active crowd.

3.2.3 Process of Crowd Behavior Analysis

The behavior of a crowd can be analyzed in one of two ways: by analyzing the behavior of each individual in the crowd or by dividing the whole crowd into small groups and analyzing each group. In medium- and high-density crowds, the former method is not applicable. Hence, the crowd is divided into groups and each group is considered a single entity to discover any changes in the environment and determine changes in behavior. Crowd behavior analysis helps us to understand the crowd dynamics, develop crowd management systems, and reduce crowd-related disasters. The challenges associated with analyzing crowd behavior using video analytics are described below:

- *Diversity of views*: Diversity of views depends on the different viewing angles of the camera. The field of view or viewing angle of a camera is the area the camera can see. The main factor that determines the viewing angle is the size of the camera lens, also known as focal length. Thus, the diversity of views depends on the focal length of the camera. The main challenge is that surveillance videos consist of a single acquisition protocol and the behavior to be analyzed is from different contexts.

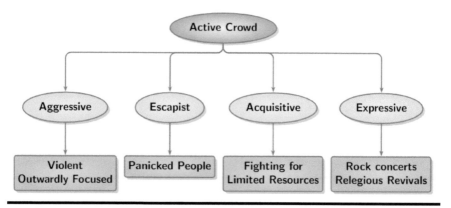

Figure 3.5   Classifications of active crowd.

■ *Occlusion in video*: Occlusion is defined as blocking the view of one object by another object. In a crowded scenario video, when human bodies overlap and walk together, occlusion happens. In dense crowds, tracking and behavior detection is a major problem because a large number of objects are in close proximity and the correspondence across consecutive frames in the video will be difficult [16]. Individual behavior analysis is easy in the case of sparse and medium-density crowds, but it is difficult in a dense crowd, due to heavy occlusions.

■ *Huge amount of data*: Since surveillance is a 24/7 process, a huge amount of data is produced. Processing this huge data is a real challenge. An appropriate computing environment is needed for the processing of this sheer volume of data.

■ *Ground truth video sequences*: The generation of ground truth video sequences for training and validation is another challenge. If such video sequences are not available, the performance of a system cannot be evaluated properly.

■ *Abnormal event classification*: Classification of anomalies in a crowded scenario is a difficult process. Since an abnormality mainly depends on the context or scenario, defining the characteristics of anomalies is a tedious process. Moreover, in real-world scenarios, anomalies should be detected from a large set of training samples, from nonexisting classes of anomalies.

■ *Low-quality video*: One of the challenges in analyzing crowd behavior is the quality of video. The main factors that affect the quality are noise and illumination variations in the video.

For crowd behavior analysis, from the captured input video the sequence of motion should be extracted. From the extracted frames, features should be identified to infer the abnormal activity. The inference can also be done from crowd models, created from the identified features. The whole process of crowd analysis is shown in Figure 3.6. A detailed discussion of the methods for each of the steps is presented in the following sections.

3.2.3.1 Motion Sequence Extraction

Smart cameras capture information in the form of video sequences. In order to analyze behavior, movement of the crowd with respect to the environment is required. Therefore, the sequence of motion is extracted from the frames of the input video to obtain the movement of the crowd. The methods of motion sequence extraction are broadly classified based on background subtraction methods, optical flow methods, and spatiotemporal methods:

■ *Background subtraction*: In the case of a stationary camera, the background is static, and the region of interest is only the foreground. Background subtraction is used to detect the foreground from the video. It is also useful

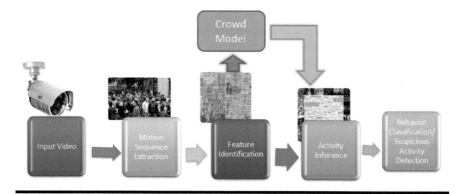

Figure 3.6   Process of crowd behavior analysis.

for detecting moving objects, such as humans and cars, from a stationary camera. The authors in [17, 18], performed background subtraction using a Gaussian mixture model (GMM) to find the presence of individuals in an isolated region. GMM uses the sum of n Gaussian components to find the probability that an observed pixel will have an intensity value I at time t. Although GMM is used for accurate foreground segmentation, it is not adaptive to varying illumination conditions. Another method for background subtraction is the center-symmetric local binary pattern (CS-LBP) [19], which is an extension of LBP. In this method, the histogram of binary patterns is computed over an eight-neighbor region. Binary patterns are created corresponding to the difference in gray-level values, from center-symmetric pairs of pixels to N equally spaced pixels on a circle of radius R. CS-LBP is used to identify the texture pattern in a local neighborhood and is helpful for dimensionality reduction. But if noise is present, the method is not reliable. Since this method uses only the difference in the spatial position of pixels, utilization of color and temporal information is not possible. In [20], a weighted moving average is used for background subtraction. In this method, the background is extracted by creating a background model, which consists of the weighted mean of pixels between successive frames. This is considered to be a robust and reliable method, but it fails in the case of a moving background.

■ *Optical flow methods*: In video data, motion is a rich source of information for performing visual tasks, such as shape acquisition, recognition of objects, and scene understanding. The aim of the optical flow estimation is to compute the velocities of the visible surface points from time-varying image intensity [21]. Kanade– Lucas–Tomasi (KLT) is the most common optical flow method for motion sequence extraction [22–24]. In KLT, a displacement vector h is calculated from the visual signals. The vector h minimizes the measure of difference between two frames $I(x + h)$ and $G(x)$

at each location *x* within a region *R*. Since the optical flow of every pixel is calculated, this is an efficient method for detecting crowd movement. But it is very sensitive to brightness change, and it is difficult to find a threshold to differentiate foreground and background pixels. Optical flow calculation of pixels can also be used for specific events, such as crowd escape detection, which uses the Bayesian model [25]. This method was used for analyzing the escape behavior of a crowd and detecting abnormalities by characterizing the presence and absence of escape situations. A conditional probability distribution function, using the Bayesian rule, is used for abnormality detection. Crowd motion for escape and nonescape conditions can be effectively modeled by this method, but it is ineffective in handling high-density crowds.

■ *Spatiotemporal methods*: Spatiotemporal computational methods are widely used when data are collected across time as well as space and have spatial and temporal properties. Since video data are rich in both spatial and temporal phenomena that exist at a certain time *t* and location *x*, we can exploit such properties for motion sequence extraction. Baumann et al. [26] proposed a spatiotemporal method called motion binary pattern (MBP) for detecting motion changes. MBP is computed in the spatiotemporal domain to gather motion information by calculating the change in pixel values across multiple frames. This method easily identifies motion and is ideal for behavioral analysis, but the motion pattern information is not very discriminatory. Local spatiotemporal motion patterns can be detected by dividing the video into cuboids [27]. For each cuboid, gradient orientation histograms are calculated from a grid of mean gradients. Each gradient orientation is quantized using regular polyhedrons. Finally, mean gradient is computed from integral video frames to extract motion patterns. This method is good for extremely crowded scenes, but it is not suitable for far-field and oblique camera setups. Spatiotemporal methods are also used to analyze stationary crowds. A three-dimensional stationary time map [28] is a spatiotemporal approach for calculating the stationary time of an object in a video. A code word is created for foreground and background pixels separately. Pixels belonging to the same code word denote that they are of the same stationary part.

When background subtraction methods are used, a new background model needs to be created each time for each set of input video. It will take time to differentiate the foreground in certain cases, such as varying illumination. Optical flow methods are time-consuming as they operate on individual pixels. Also, they spend their search time within intensity regions that do not have any motion. As spatiotemporal methods utilize space and time information for motion extraction, the limitations of the other two methods are rectified. After extracting the motion, features are identified for further classification and prediction.

3.2.3.2 Feature Identification

Features are specific characteristics that help to classify or detect events from videos. In the case of a crowd, the main features are pixel gradient descriptors, edge pixels, trajectories, blobs, locations of key points that help to identify humans, forces between individuals, stationary time of the pixel, and so on. The features can be divided into local and global/holistic based on the approaches used to extract the same.

- *Local features*: Local features are referred to as a distinct structure in the video frame that differs from its surroundings. The methods for extracting local features are summarized in Table 3.2.

  One of the common feature extraction methods is scale-invariant feature transform (SIFT) [29–31]. SIFT helps to detect the location of interest points as maxima or minima of the difference of Gaussians (DoG) in scale space. SIFT uses the scale space extrema detection by DoG for locating the potential key points. The key points are then localized by Taylor series expansion of the scale space, and the edges are removed using a Harris corner detector. After this step, only the strong interest points remain. Then orientation histograms are taken of the interest points to create key points with the same location and scale. Finally, the key point descriptor is found by taking the orientation histogram of the 16 ×16 neighborhood around the key point, which is represented as a vector. SIFT is ideal for feature matching, but it fails in complex scenarios such as dense crowds, as the number of objects in the scene is very large.

Table 3.2 Local Features

Methodology	Features
Scale-invariant feature transform (SIFT)	Good for feature matching
Histogram of oriented gradient (HOG)	Used for object detection
Motion boundary histogram (MBH)	Robust for camera motion
Deformable part model (DPM)	Detects humans in images
Spatiotemporal gradients	Works well for low-resolution scenarios
Histogram of optical flow (HOF)	Good for the detection of moving crowds
Stack convolutional ISA	Learns features directly from video data

Histogram of oriented gradient (HOG) is another commonly used local feature extraction method [20,32–34]. In HOG, the video frames are divided into spatiotemporal patches. Each patch is divided into cells and histogram descriptors are calculated. The histograms are normalized and coupled together to obtain the final descriptor, which consists of edge orientations as histogram bins. HOG features are usually used to detect humans in surveillance systems. But in highly occluded scenarios, HOG tends to fail as it detects nonhumans as humans. In [35], a method called motion boundary histogram (MBH) was used to extract the features. MBH divides the video frame into x and y components that split the optical flow field. For each x and y component, spatial derivatives are computed separately. A histogram of the components is also quantified to encode the orientation information. MBH gives better results in situations where the background is dynamic. But it is sensitive to localization and tracking errors.

A method for extracting features by detecting humans from surveillance video is proposed in [33] and [36]; it is known as the deformable part model (DPM). Features are extracted from objects by considering them as parts. Using models, the detector first finds a match of its whole, and then it uses the part models for finer results. The deformation model consists of a root filter and a set of part filters. This detector uses a sliding window approach, where the root filter is applied at all positions and the frame is scaled. Human detection using DPM is very reliable in the case of still images. But in videos, it cannot differentiate scene-specific information from frames, such as the color pattern of objects and the background. Local motion features can be extracted by calculating the spatial and temporal gradients [37]. Here, initially a spatial edge is calculated using the Sobel edge detector. Then gradients are calculated over the spatial edge with respect to the time difference between the current spatial edge image and the moving average of the edge image. This method works fine in low-resolution scenarios, but it is not effective when noise is present.

The authors of [38] proposed a method called histogram of optical flow (HOF) for feature identification. The motion information, such as orientation, magnitude, and entropy, is encoded in a HOF descriptor. From the information, optical flow is calculated, and it is quantized into histogram bins. As this method explores the flow of motion, moving crowd detection is effortless. A deep learned local descriptor for feature identification using stack convolutional independent subspace analysis (ISA) is employed in [39]. Stacked ISA is a two-layered network where the first layer is composed of multiple ISAs. The input of the first layer is extracted as cuboids, directly from the video sequence. Before giving the combined output of the first layer to the second one, the dimension is reduced by applying principal component analysis (PCA). All the local feature extraction methods discussed above focus on adapting features from static images to the video domain. However, stack convolutional ISA is an unsupervised method for learning the features directly from video data. Extraction of low-level visual features is pragmatic in crowd behavior analysis, as it seeks to detect small intensity variations in videos to identify an event.

■ *Global/holistic features*: Global/holistic feature descriptors extract the features by considering the video frame as a whole, to generalize an entire object as a vector. Kaltsa et al. [20] explore the properties of swarm intelligence to extract features using the histogram of oriented swarms (HOS). In HOS, the motion features are extracted by modeling the optical flow values using the concept of swarms. The optical flow magnitude values of each pixel across multiple frames are used to construct a swarm histogram. The swarm is considered an agent, and optical flow values are assumed to be its prey. This is an efficient method to capture motion features in the midst of noise and occlusion, when combined with the other local feature extraction methods. Global features can also be extracted using edge segmentation [24], by splitting the frames into local coherent regions. Edge segmentation starts with the canny edge detection method, and the pixel values in the edge image are rearranged in the form of a vector, which is given as input to a Delaunay triangulation phase. This method is suitable for indoor environments, but it fails in cases of complicated backgrounds. Moreover, the foreground and background segmentation in a crowded scenario is challenging. In [27], hidden Markov model (HMM) is used to extract the features from a crowded scenario. HMM is used to excerpt the temporal information in a video sequence from a spatial cuboid of the frame. Since a feature has to be extracted from physically moving objects, cuboids in the same spatial location show the Markov property in the temporal domain. The Markov property is described with a set of hidden states and the transition from one state to another. The state information in the spatial cuboids is used to model the Markov chain for identifying the features.

Suspicious activity detection and behavior analysis in a crowded situation works well only with the identification of good features. The combination of multiple local features or the fusion of local and global features can be applied, based on the context to get more desirable results.

3.2.3.3 Methods for Analyzing Crowd Behavior

Crowd behavior analysis from video data is an active research area in the field of computer vision. Due to the growth in the human population, the formation of crowds in public areas is increasing. Nowadays, crowd-related disasters are common. Therefore, effective and efficient prediction of disasters is indispensable to avoid crowd disasters. For that purpose, the behavior of the crowd should be analyzed accurately. The methods used for behavior analysis in literature are classified based on crowd emotions, physics, trajectory extraction, neural networks, and deep learning.

- *Crowd emotions*: The emotion of a crowd is an important factor that corresponds to the behavioral changes of the crowd. The authors of [39] proposed a bio-inspired model for detecting crowd emotions. They used a dynamic probabilistic mechanism to predict the behavior based on crowd emotion. Rabiee et al. [40] proposed an attribute-based approach to analyze crowd behavior, where the attribute is the crowd emotion. They used a k-dimensional binary vector to represent emotions such as anger, sadness, happiness, excitement, fear, and neutral. A linear support vector machine (SVM) is used for emotion classification. In [41], crowd mood is determined by inspecting the spacing interactions and structural levels of motion patterns. The crowd mood is represented by structured trajectory learning and arousal-valence motion features. Finally, the crowd mood is modeled using AdaBoost feature weight selection and emotion state regression by plotting the crowd mood curve.
- *Physics*: A moving crowd can be analyzed by using aspects of physics concepts such as particle force and fluid dynamics. Most of the methods in the literature use the social force mechanism to identify crowd dynamics and patterns. Social force uses the interaction between individuals to diagnose crowd behavior. It is an application of particle dynamics where each object is treated as a particle and the interaction between them is simulated as forces. Newton's equation is then applied to compute the acceleration, velocity, and position from the summed-up particle forces [42]. In addition to the computation of displacement velocity and acceleration, the histogram of oriented pressure can also be computed from the social force model by combining the interaction force between objects [43]. This model requires prior knowledge of the scenario, and it only considers the temporal characteristics of the data. The force between particles or objects can also be consolidated as an energy function to analyze the crowd behavior [44]. A structural context descriptor (SCD) is created by adopting the potential energy function of the interforce between particles for interlacing the visual context. By investigating the SCD variation of the crowd, the relationship among individuals is identified to localize the abnormality among them. The repulsive force between particles is also used to classify the behavior of a crowd [45]. Here each individual is represented as three overlapping circles, such as the head, upper body, and lower body. The psychological and physical forces of these parts to form subgroups are represented as a discrete element model (DEM) for further analysis. The concept of fluid dynamics is also explored in crowd behavior analysis. The individuals are considered particles in fluid, and crowd motion is shaped based on the lattice Boltzmann model [46]. The crowd particle collision and particle direction are analyzed to calculate the velocity of the crowd. Abnormal events are identified through particle streaming and collisions.

■ *Trajectory extraction*: A trajectory is the direction of motion of an object that follows through space as a function of time. Crowd trajectories are extracted for analyzing the direction of motion, and thereby behavior can be determined. In [46], trajectories are extracted using the properties of particle hydrodynamics. Individuals in a crowd are considered particles, and those with the same velocity follow the same direction. The presence of abnormality is identified by verifying the presence of collision between particles. Rabiee et al. [40]proposed the dense trajectory method to analyze human behavior. The dense trajectory is calculated from a bag of features codebook, generated from HOG, HOF, MBH, and the trajectory. Euclidean distance is calculated to find the closest vocabulary for each video descriptor. Another method for representing the trajectory data using the activity description vector (ADV) was proposed in [23]. Here, ADV is calculated by determining the number of occurrences of a person in a specific point of time. The local movements are then classified using clustering methods to analyze the behavior. As the motion analysis of trajectory extraction methods focuses on each subject individually, they are applicable for situations with few moving objects.

■ *Artificial neural networks (ANNs)*: ANNs are computing systems that replicate a biological neuron to learn tasks through examples. The authors of [40, 47] use ANN to classify crowd behavior as normal or abnormal. The characteristics of ANN mainly depend on the training data, which fail in real-time environments where there are scenarios that are not present in the training samples.

■ *Deep learning*: Deep learning is an area of machine learning in which algorithms and structures are based on a hierarchical form of ANNs. Deep learned visual descriptors using L_2 normalization were used to create codebook vectors that can be used to analyze crowd behavior [39]. Shao et al. [48] proposed a deep model that uses crowd motion channels that consist of properties such as appearance and motion. A deep convolutional network was proposed in [49] for crowd density estimation; it takes image patches from training images as input. In general, deep learning models have the ability to perform automatic feature extraction from raw data, which is very useful in analyzing crowd behavior.

Crowd behavior analysis using deep learning approaches is considered to be more efficient as the smart surveillance environment needs to adapt to the real-time situation without adequate training samples. Effective and efficient behavior analysis is essential for the accurate identification of suspicious activities in a crowd. Any behavior that is bizarre in occurrence and deviates from normally understood behavior can be classified as suspicious. There are many papers in literature that identify the presence of suspicious or abnormal activities. In [44], abnormality is detected from the motion difference between individuals, computed using the

HOF. This method works with a medium crowd to classify it as normal or abnormal and is not meant for dense crowds. A lattice Boltzmann–based crowd anomaly detection proposed by Xue et al. [46] also classifies the crowd as normal or abnormal, based on the crowd trajectories. In short, they attempt to classify normal or abnormal trajectories of individuals in the crowd. An anomaly detection algorithm developed by Wang and Xu [50] uses spatiotemporal texture features to classify the crowd as normal or abnormal. Sabokrou et al. [51]proposed a deep fully convolutional neural network to classify crowd behavior as normal or suspicious or abnormal. The suspicious behavior is further processed to classify it as either normal or abnormal. The problem with the papers in the literature is that detection of suspicious activity is classified as either normal or abnormal, which is mostly ambiguous on the semantic level.

But an IoT-based smart surveillance environment always demands recognition of the type of suspicious activity and its situation to make proactive decisions. Moreover, there are a wide variety of sensors (camera, visual sensors, and mobile devices) in smart surveillance surroundings that comprise data from different modalities other than video (video data from smart cameras and visual sensors). Therefore, to analyze crowd behavior and detect suspicious activities, knowledge of the different sensor data analytics methods is necessary. Hence, the next section demonstrates the data analysis techniques from sensors.

3.3 Sensor Data Analytics

In an IoT-based smart surveillance system, in addition to cameras that consists of visual sensors that produce video data, other type of sensors can be integrated. Video-based behavior analysis uses videos or images from cameras, while sensor-based analysis focuses on the data from smart sensors, such as accelerometers, gyroscopes, and sound sensors [52]. The types of sensors that can be used for behavioral analysis are discussed below.

- *Motion sensors*: Motion sensors are used to detect moving objects, mainly people. These can be incorporated as a component of a surveillance system to alert motion. The most commonly used motion sensors are (1) passive infrared (PIR) sensors, which are sensitive to the temperature of a person's skin and use midinfrared wavelengths to detect motion; (2) microwave sensors, which emit continuous microwave radiations (the reflection of the rays from the objects is calculated to detect the motion); (3) ultrasonic sensors, which use ultrasonic waves to detect the movement of objects; (4) tomographic sensors, which are used to create a mesh network of radio waves and detect motion when a disturbance in the network is identified; and (5) combined types, which are combinations of the above four sensors that are used to detect motion to increase efficiency.

- *Pressure sensors*: Pressure sensors work using the variation of capacity, variation of resistance, piezoelectricity, and so forth. The behavior of a crowd can be analyzed by using a set of pressure sensor data based on the variation of resistance. In this case, the pressure is converted to electrical signals [53].
- *Acoustic sensors*: Acoustic sensors are used to detect sound and audio in a specified area. The three acoustic entities used for crowd monitoring are sound intensity, sound pressure, and sound loudness [54].
- *Microelectromechanical systems (MEMSs)*: These devices are built by means of integration of miniaturized mechanical and electronic equipment through microfabrication technology. Most of the body-worn sensors are MEMS devices. One of the most common MEMS sensors is the accelerometer. It is a device used to measure the acceleration of the body. Another common device is the gyroscope, which is used to measure the angular velocity of an object and identify the flow pattern in a crowded scenario. Using the accelerometer and gyroscope, one can measure the change in acceleration and angular velocity of human movement. They are often found in smart phones, watches, glasses, and helmets [52].

Crowd behavior analysis and suspicious activity detection are possible by integrating different types of sensors. With this integration, there is an exponential increase in the data. More advanced methods and algorithms are needed to process the big data. The entire process of sensor data analytics is shown in Figure 3.7. The process starts with the data acquisition and cleaning of raw data from multiple sensors. Then, by exploring different analytics algorithms, the data are modeled and analyzed based on the context before being used for event prediction and visualization. Different tools and platforms are available in the literature for the entire process.

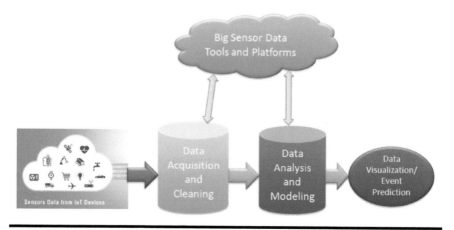

Figure 3.7   Process of sensor data analytics.

The challenges associated with data processing are listed below, before going to a detailed discussion of the entire sensor data analytics process.

- ▪ The data are coming from different sources and are inherently noisy.
- ▪ Due to the malfunctioning of sensors or malicious network attacks from external parties, the measurements of data may lie outside the expected range. This type of data is known as outlier data.
- ▪ The size of the data is very large.
- ▪ Since different types of sensors are used, data from multiple sources may lack context or be incomplete.
- ▪ There may not be a computing environment to process the sheer volume of data.

Extracting useful information from raw sensor data is a trivial task. This knowledge discovery process can be done by effective mining of the data. Data mining is used to discover subtle patterns among the data; it is necessary in modeling relationships among sensor measurements. Before applying mining, the extracted data should be cleaned to address some of the challenges, such as noise, outliers, missing data, or data from malfunctioning sensors, and data reduction in the case of real-time data.

3.3.1 Data Acquisition and Cleaning

For sensor data analysis, the first step is to extract and clean the data. This involves identifying and generating the required data from multiple sensors and obtaining clean and meaningful data. Since the sensor-acquired data are often noisy, this step is very important to remove the uncertainties. In sensor data, there is a high degree of spatiotemporal correlations. Multiple measurements are needed to reflect the information about an event due to the overlapping areas of coverage. Thus, there is an unpredictability among multiple sensor measurements that affects the quality of the data [55]. The methods used for data cleaning are described below and are also summarized in Table 3.3.

- ▪ *Noise removal and filtering*: Since the data are coming from different sources, acquired data are often noisy. In [56], spatiotemporal regression and a Kalman filter is used for cleaning data.
  - A Kalman filter is used to determine the state of a dynamic system from incomplete and noisy measurements using a recursive process. It is based on linear algebra and the HMM. The algorithm can run in real time with present input measurements and previous state without any additional information about the past. The two different phases are called the predict phase, which is used to estimate the current step from the previous time step, and the update phase, which is used to refine the prediction to a more accurate estimate.

Table 3.3   Methods for Data Cleaning

Purpose	Method	Features
Noise removal and filtering	Kalman filter	Used to filter out noisy data Can run in real time
	Spatiotemporal regression	Extracts the best set of data values by using regression
	Empirical mode decomposition	Filters the high-frequency data for denoising
Removal of outlier and redundant data	Multivariate Gaussian model	Identifies outlier data Detects the sensor that generates erroneous data
	kernelPCA	Filters out redundant data Reduces the dimensions of data
	Bayesian belief network (BBN)	Classifier for outlier data detection Classified based on probability
Missing data generation	Spatial and temporal interpolation	Generates both spatial and temporal missing data

- Spatiotemporal regression is the process of fitting a curve to the best set of real values. The sensor data consist of two components: space and time. The time dependency of the data is captured by fitting a time polynomial by using weights and basis functions. In the case of sensors, where data are streaming, a sliding window approach is used to extract the data needed for regression [56].
- Empirical mode decomposition is a method for sensor data noise filtering presented in [57]. The raw data are decomposed into several intrinsic mode functions (IMFs). The data are reconstructed by eliminating the first IMF where high frequencies are filtered out.

■ *Removal of outlier and redundant data*: As a result of sensor malfunctioning, some data may lie outside the expected range and are called outlier data. Zhang et al. [58] used a model called the multivariate Gaussian model to detect and identify errors from data retrieved from a group of sensors. Outliers can also be detected and cleaned by modeling sensor data using Bayesian belief networks(BBNs) [59].

- The multivariate Gaussian model is used to find the correlated data changes from a group of sensors. This method is used to find outlier data detection. The model consists of two steps: dirty data detection and sensor error identification. Dirty data detection is used to detect the outlier data points. After detecting erroneous data, the specific sensor that produces the data can be detected using the second step in multivariate Gaussian distribution—the sensor error identification method.

– Bayesian belief network: A BBN is a directed acyclic graph with nodes representing random variables and edges representing probabilistic dependence between two variables. Using BBN, one can compute the joint probability distribution of variables as the product of local distributions between a node and its parents. In [59], BBN is used as a classifier for outlier detection. Using belief propagation, the sensor base station compares the probability of its most likely data with the probability of its actual sensed reading. If the difference is large, the data are considered as outlier.

As a result of the sparse geometrical relationship between the sensor nodes, redundant values are also present in the original sensor data. To avoid this sensor network error, erroneous data should be filtered out from raw data. A method using PCA was proposed by Zhang et al. [58]. They use the kernelPCA approach to remove the redundant data, and as a result, data dimension is also reduced.

– In kernelPCA, a kernel is applied to map the raw data to the feature space before applying the PCA. In [58], the kernel function is designed based on the topology of sensor nodes. Every sensor node has the capability to detect the received signal strength and direction of the arrival angle. From this information, a feature space is created by using the time difference, and a covariance matrix is calculated to find the m principal components. The principal components are matched to a linear vector space and error value is calculated between train data and test data. If the error value is less than a specified threshold, then all sensors are good.

■ *Missing data generation*: In certain cases, if the sensors are not well positioned, then there exist missing data points in both space and time dimensions. The authors of [60] proposed interpolation of data in both the spatial and temporal dimensions, to take care of the missing data. Spatial interpolation is used to fill the missing data between two measurements at time t. Data stored in the same time interval are interpolated using k-nearest-neighbor interpolation, based on Manhattan distance. Temporal interpolation is used to generate missing data between two long time intervals. In this case, interpolation is done with values of the previous time intervals.

Data cleaning aims to correct the errors associated with data extraction. The cleaned data are given for further processing analytically or statistically. The next section discusses the analytical and statistical methods in the literature to model the data.

3.3.2 Analysis and Modeling of Data

Data analysis and modeling is used to provide an accurate and deep understanding of data through analytical or statistical methods. The methods can be broadly classified based on the approaches used in analyzing the data. The different approaches in literature are summarized based on deep neural networks, machine learning, ensemble systems, graph models, and statistical analysis.

- *Deep neural networks*: A deep neural network is a computational model that consists of interconnected nodes organized in many vertical layers. A deep network can process highly varying patterns and build an efficient feature space [61]. To perform $PM_{2.5}$ (particulate matter) concentration predictions in Japan, data collected from sensors are trained using deep forward neural network and deep recurrent neural network. Also, for time-series prediction, an autoencoder is used for pretraining [61]. By utilizing GPS data from taxies, traffic congestion prediction was done in [62]. Here, deep restricted Boltzmann machine and recurrent neural network architecture were used to model and predict traffic congestion evolution. Lv et al. [63] proposed a traffic flow monitoring system using a stacked autoencoder. The data for the prediction were collected from Caltrans Performance Measurement System (PeMS) database, which consists of the data from detectors deployed in freeway systems across California. A greedy layer-wise unsupervised learning algorithm was also used to pretrain the deep network to improve the prediction performance.

- *Machine learning*: Algorithms and methods that can be used for learning and making predictions of data are called machine learning algorithms. The ability to effectively learn and predict data has made machine learning algorithms popular. An intelligent system for predicting human behavior in the case of natural disasters is proposed by the authors in [64]. Here, probabilistic reasoning is used for the prediction of evacuation and movement of people based on GPS data. The probabilistic inference for evacuation is calculated using Bayes' rule, and posterior probability is required for prediction. For predicting air quality, a method was proposed by Zheng et al. [65], which uses a semisupervised learning approach, based on two classifiers. Spatial correlation of the data is classified using ANN [57], and temporal features are classified using a linear chain conditional random field (CRF). The method proposed by Xia et al. [66] also employs a back-propagation ANN to classify data for the estimation of the presence of $PM_{2.5}$ in the atmosphere.

- *Ensemble systems*: Ensemble systems are used to learn information from several sources, where there are heterogeneous features and a single classifier is not sufficient. Robi Polikar [67] proposed an ensemble system that uses multiple machine learning algorithms and classifiers. In this approach, ensemble generation techniques, such as bagging, AdaBoost, and mixture of experts, and classifier combination strategies, such as algebraic combiners, voting methods, and decision templates, are used.

■ *Graph models*: In [68], a graph-based model is created to analyze the environmental data to study the impact of burning fuel in New York City. In this case, a weighted directed graph is created by considering each data element as a node. Each data element in a data set is considered a building. The edges are defined using a modified Gaussian dispersion plume model. If there is a self-loop, it measures the emission of a building *i*. Indegree and outdegree are used for detecting the relative emission of polluted air between buildings.

■ *Statistical analysis*: One way of performing statistical analysis of data is by using the homoscedasticity measure. It is the detection of novelties in sensor signal, in case of a sudden signal change, due to external factors. In [57], Leven's test is performed to find this measure. For determining self-similarity, a statistical measure called autocorrelation is used in sensor data [57]. It is the measure of self-similarity calculated from an autocorrelation function at different time periods. Entropy measure, coefficient of variation, and minimum ratio are the other most common statistical measures.

After choosing the correct algorithms and methods for data analysis, the next step is to identify the tools and platforms for data processing. The following section summarizes the current tools and platforms for IoT data processing.

3.3.3 Tools and Platforms for Data Analysis

This section explains the data processing and analytics platforms that can be used for large amounts of sensor data. The data processing and analytics can be done closer to the data source using computing environments such as mobile edge computing, cloud computing, and fog computing [69].

■ *Apache Hadoop*: Hadoop is an open-source data processing platform that can store and process large amounts of data. It consists of a distributed file system called Hadoop Distributed File System (HDFS) for storing the data and dividing the data into smaller parts. Another component is MapReduce, which helps in the parallel processing of a large set of data in a distributed manner [3,70–72]. The major advantages are its scalability, cost-effectiveness, flexibility, and speed. The main limitation of this tool is the absence of encryption at the storage and network layer [73].

■ *Kafka*: Kafka is a stream analysis tool with high throughput and low latency for real-time data. Kafka consists of components that can be defined based on the context. The components are called producers and consumers. For example, in traffic management, one type of message stream consists of the speed of a vehicle. Vehicles (producers) publish messages to traffic police (consumers) to subscribe and pull new messages when they become available [71]. Kafka is mainly used for data-driven applications.

■ *Mahout*: Mahout is a distributed machine learning framework that can be used on Hadoop [70,72]. The algorithms in Mahout can be efficiently used for clustering and classification.

■ *NoSQL database*: NoSQL helps in the easy storage and retrieval of data as the data are stored in a tabular form, similar to a relational database [71].

■ *Spark*: Spark is an open-source, fast, and general platform for large-scale big data processing [71,72]. It is a cluster computing architecture that provides streaming analytics and a distributed machine learning library for feature extraction, sampling, and so forth.

■ *Hive*: Hive is a data warehouse and data analysis software framework on the Hadoop platform [71,72]. Hive provides data summarization, query analysis, and data processing.

■ *Pig*: Pig is a platform used for parallel data flow execution on Hadoop [71,72]. It is used for processing both structured and unstructured data.

■ *Xively*: Xively is an online open system that enables users to submit sensor data and build applications based on that data. It also helps to embed real-time graphs, analyze and process historical data, and send real-time alerts [74].

■ *Cloudera data hub*: Cloudera is a Hadoop-based framework for processing sensor data. This tool can handle massive amounts of data and consists of various components, such as the Cloudera Navigator, for data discovery and metadata management [73].

■ *WikiSensing*: This is an online database service that helps to connect sensor devices and store the data. Developers can connect to this database and build their own applications. It also helps to annotate information about the sensors and their data [74].

■ *1010data*: 1010data is a database that has the capability to deal with IoT data. This tool can be used for optimization, statistical analysis, visualization, and data integration. It provides advanced analytic capabilities as well as access control for accessing back-end systems [73].

Other tools include SAP-Hana, which provides distributed solutions; HP-Haven, which uses a high-performance computing environment; Hortonworks, an open-source software that works on the Hadoop platform; and the Pivotal Big data suite, a cloud-based tool that helps in parallel processing [73].

Even though many data analytic methods are discussed in the literature for predicting air quality, for determining traffic flow, none of them point out the use of sensor data in analyzing crowd behavior. By deploying sensors like accelerometers and acoustic sensors in a crowded place, the analysis and prediction of crowd behavior become effortless. In addition to sensors and video, mobile phones can also contribute data, as today's smart phones are outfitted with sensors, which is demonstrated in the next section.

3.4 Mobile Crowd Sensing and Behavior Analysis

The method in which a large group of persons possessing mobile devices with sensing capabilities share data for specific purposes is referred to as mobile crowd sensing. Data acquisition from mobile sensors possesses many advantages. First, today's smart phone devices have more computing and communication resources, along with heterogeneous sensing capabilities. This will help in many applications that require resources, especially in a crowded scene. Second, we can frame large-scale sensing applications easily as each person is carrying mobile devices wherever they go. For example, traffic data collection from cameras and sensors for detecting congestion can be replaced with data from smart phones carried by drivers [75]. So mobile crowd sensing data should be exploited for the prediction of disasters in the case of crowd analysis as well.

Few papers are available in the literature for crowd management utilizing the data from sensors and smart phones. The authors in [3] proposed an architecture for crowd sensing or tracking from big data. Here crowd tracking in a smart city is done with the help of Wi-Fi scanners from mobile phones. The scanners gather the data about the movement of crowds. These data are preprocessed by cleaning unwanted data and then continuous flow is analyzed using data correlation with historical events. Crowd sensing as a service using a cloud platform is proposed in [76] for crowd management. In this approach, to avoid disinformation from malicious users and false sensor readings, trustworthy sensing for reputed crowd management is presented by collecting data from smart phones based on an auction mechanism. Another crowd management system proposed by Franke et al. [76] also uses mobile phone sensing for location and information retrieval. The authors created a crowd management application for providing event-specific information to users via mobile phones, collecting sensor data from users, and translating the data to represent the crowd state, instruct people to act differently in case of an emergency, and analyze important events for future planning. The authors in [77] proposed a framework for emergency management in mass gatherings, based on crowd emotions. The data for this framework are collected from social media messages to determine how people feel and think. An emotion model of the crowd is created using rule-based reasoning and a bag of words.

Mobile crowd sensing is also used as a sensing platform for smart cities [78] that leverages IoT technology to sense the weather, environment, public transport, and people flow. In [78], the passenger flows in the city of Porto, Portugal, are collected as metadata from Wi-Fi connections of passengers in public buses. This will help to improve the public bus service by understanding the flows of passengers. In a recent work by Bellavista et al. [79], mobile crowd sensing is exploited to support the effective deployment of human-driven mobile edge computing. It uses human social and mobility effects to extend mobile edge computing coverage. Mobile crowd sensing helps to continuously monitor humans and their mobility patterns to dynamically reidentify locations of potential interest for the deployment of new edges.

In short, mobile crowd sensing uses spatiotemporal patterns to analyze data from a group of mobile devices [75]. In the case of crowd behavioral analysis, such knowledge of the spatial distribution of individuals is indispensable to analyze their movements. Since the spatial patterns are disseminated over various timescales, temporal information is also accessible. This insight helps in the better coordination and behavior analysis of crowds, depending on the time of day, and it also helps in long-term planning for avoiding crowd disasters and suspicious activities. For example, individuals can report problems if any suspicious activity is about to happen. In such cases, it will be convenient for the officials to make proactive decisions.

According to the postulates presented by Martella et al. [80], crowd management and effective crowd analysis are possible only by incorporating the new developments in technology, including IoT devices such as sensors (visual, motion, acoustic, and mobile sensors), decision support systems, and sophisticated communication and interaction paradigms.

3.5 Conclusion and Future Perspectives

With the increasing human population, the formation of crowds in public places is inexorable. As a result, crowd-related crimes and disasters are common. So, effective analysis of surveillance data from crowded places is ineluctable to prevent such incidents. In this chapter, we provided an extensive review on the analysis of IoT data (from video cameras, sensors, and mobile devices) for analyzing crowd behavior. Behavior analysis of an individual is possible in sparse and medium-density crowds but may fail in an extremely dense crowd. This is because in dense crowds a large number of objects are in close proximity, which makes it difficult to analyze persons. Therefore, the monitoring of crowded scenes requires the monitoring of an excessive number of individuals and their activities. In this context, data from multiple modalities, such as video, sensor, and mobile devices, can be integrated to achieve the objective. The chapter uncovers a wide variety of analytics methods for determining the behavior of the crowd in this chapter. The outcome of this review demonstrates the need for algorithms and models to classify crowd behavior from multimodal data. The future perspectives are summarized in the following paragraphs.

A group of people that are not in the moving state is called a stationary crowd. The detection of stationary groups has important applications in a smart surveillance environment because the sudden evolution and dissemination of stationary groups might lead to variations in crowd patterns. Individuals present at a scene for a longer time may often lead to security issues. The audiences in sports arenas and political protests are considered stationary. The behavioral analysis of such crowds is important, as more abnormal activities may take place in connection with such crowds. Almost all papers in the literature analyze the behavior of crowds based on their motion from video data. In addition to video data, stationary people with mobile devices also provide important information to analyze such crowds. This entails the need for combining data from visual sensors with mobile crowd sensed data to detect the crowd behavior.

The detection of suspicious activity in a smart surveillance system suffers from the impracticality of detecting anomalies from nonexisting classes, because normal and abnormal events are obscure at the semantic level. Certain crowd behaviors are normal in one context but suspicious in others. For example, crowds participating in a rally are normal, but people suddenly starting to run in an airport indicates an unexpected incident. Therefore, the surveillance system should differentiate the anomaly based on the behavior of the crowd. Thus, a thorough analysis of different modalities of data is significant to classify suspicious activities. Feedback from common people in the form of mobile crowd sensing can also be exploited to analyze the situation.

The amount of data generated by smart surveillance devices is growing exponentially as the devices work nonstop, producing torrents of data that are unmanageable by current data processing and analytics techniques. Present-day IoT devices utilize cloud platforms for computation purposes, which suffer the limitation of low-latency response for applications that use data from nearby sources. Moreover, moving all data generated from the IoT devices to the cloud server involves more storage and Internet infrastructure. This is often expensive, impractical, and irrelevant. The above shortfalls of the cloud can be overcome by employing the fog computing paradigm. Fog creates a set of microclouds or fog nodes near the sources of data. Fog nodes process the data before they reach the cloud. This will reduce the transmission time and cost and thus reduce the need for huge data storage. It connects the IoT devices and cloud data centers by facilitating the use of storage and network services closer to the devices. Surveillance data analytics with fog computing is of predominant importance as it provides efficient knowledge discovery and smart decision support.

References

1. G. L. Foresti, C. Micheloni, C. Piciarelli, and L. Snidaro, "Visual sensor technology for advanced surveillance systems: Historical view, technological aspects and research activities in Italy," *Sensors*, vol. 9, no. 4, pp. 2252–2270, 2009.
2. "List of human stampedes" [Online]. Available https://en.wikipedia.org/wiki/List_of_human_stampedes [accessed 2 January 2018].
3. C. Chilipirea, A.-C. Petre, L.-M. Groza, C. Dobre, and F. Pop, "An integrated architecture for future studies in data processing for smart cities," *Microprocess. Microsyst.*, vol. 52, pp. 335–342, 2017.
4. Y. L. Tian, L. Brown, A. Hampapur, M. Lu, A. Senior, and C. F. Shu, "IBM smart surveillance system (S3): Event based video surveillance system with an open and extensible framework," *Mach. Vis. Appl.*, vol. 19, no. 5–6, pp. 315–327, 2008.

5. J. Fernández et al., "An intelligent surveillance platform for large metropolitan areas with dense sensor deployment," *Sensors (Switzerland)*, vol. 13, no. 6, pp. 7414–7442, 2013.

6. R. Baran, T. Rusc, and P. Fornalski, "A smart camera for the surveillance of vehicles in intelligent transportation systems," *Multimed. Tools Appl.*, vol. 75, no. 17, pp. 10471–10493, 2016.

7. D. Eigenraam and L. J. M. Rothkrantz, "A smart surveillance system of distributed smart multi cameras modelled as agents," in *2016 Smart Cities Symposium Prague (SCSP), 2016*, pp. 1–6, 2016.

8. "Bosch intelligent video analysis" [Online]. Available http://resource.boschsecurity.com/documents/Commercial_Brochure_enUS_1558886539.pdf [accessed 2 January 2018].

9. "Bhubaneswar's 'Smart Safety' city surveillance project powered by Honeywell Tec hnologies" [Online]. Available https://www.honeywell.com/newsroom/news/2015/05/bhubaneswars-smart-safety-city-surveillance-project-powered-by-honeywell-technologies [accessed 5 January 2018].

10. Hitachi, "Data integration helps smart cities fight crime," IoT-Hitachi-Smart Communities-Solution, pp. 1–4, 2015. Available https://www.intel.com/content/dam/www/public/emea/xe/en/documents/solution-briefs/iot-hitachi-smart-communities-solution-brief.pdf [accessed 10 July 2018].

11. S. Talari, M. Shafie-khah, P. Siano, V. Loia, A. Tommasetti, and J. Catalão, "A review of smart cities based on the Internet of things concept," *Energies*, vol. 10, no. 4, p. 421, 2017.

12. "iOmniscient" [Online]. Available https://www.intel.com/content/dam/www/public/emea/xe/en/documents/solution-briefs/iot-hitachi-smart-communities-solution-brief.pdf [accessed 5 January 2018].

13. S. Miyazaki, H. Miyano, H. Ikeda, and R. Oami, "New congestion estimation system based on the 'crowd behavior analysis technology,'" *NEC Tech. J.*, vol. 9, no. 1, pp. 78–81, 2015.

14. B. Yogameena and C. Nagananthini, "Computer vision based crowd disaster avoidance system: A survey," *Int. J. Disaster Risk Reduct.*, vol. 22, pp. 95–129, 2017.

15. S. Ali and M. Shah, "Floor fields for tracking in high density crowd scenes," *Lect. Notes Comput. Sci. (including Subser. Lect. Notes Artif. Intell. Lect. Notes Bioinformatics)*, vol. 5303 LNCS, part 2, pp. 1–14, 2008.

16. M. Kumar and C. Bhatnagar, "Crowd behavior recognition using hybrid tracking model and genetic algorithm enabled neural network," *Int. J. Comput. Intell. Syst.*, vol. 10, pp. 234–246, 2017.

17. D. Y. Chen and P. C. Huang, "Visual-based human crowds behavior analysis based on graph modeling and matching," *IEEE Sens. J.*, vol. 13, no. 6, pp. 2129–2138, 2013.

18. B. Yogameena and K. Sindhu Priya, "Synoptic video based human crowd behavior analysis for forensic video surveillance," in *Advances in Pattern Recognition (ICAPR), 8th International Conference, IEEE*, pp. 1–6, Kolkata, India, 2015.

19. M. Heikkilä, M. Pietikäinen, and C. Schmid, "Description of interest regions with center-symmetric local binary patterns," *Comput. Vision Graph. Image Process.*, vol. 2, pp. 58–69, 2006.

20. V. Kaltsa, A. Briassouli, I. Kompatsiaris, L. J. Hadjileontiadis, and M. G. Strintzis, "Swarm intelligence for detecting interesting events in crowded environments," *IEEE Trans. Image Process.*, vol. 24, no. 7, pp. 2153–2166, Springer, Boston, MA, 2015.

21. D. Fleet and Y. Weiss, "Optical flow estimation," in *Handbook of Mathematical Models in Computer Vision*, pp. 239–257, Springer, Boston, MA, 2005.

22. M. S. Zitouni, H. Bhaskar, J. Dias, and M. E. Al-Mualla, "Advances and trends in visual crowd analysis: A systematic survey and evaluation of crowd modelling techniques," *Neurocomputing*, vol. 186, pp. 139–159, 2016.

23. J. Azorín-López, M. Saval-Calvo, A. Fuster-Guilló, and J. Garcia-Rodriguez, "Human behaviour recognition based on trajectory analysis using neural networks," in *Neural Networks (IJCNN), The IEEE International Joint Conference on Neural Networks (IJCNN)*, pp. 1–7, Dallas, TX, 2013.

24. A. Pennisi, D. D. Bloisi, and L. Iocchi, "Online real-time crowd behavior detection in video sequences," *Comput. Vis. Image Underst.*, vol. 144, pp. 166–176, 2015.

25. Z. Yu, S. Wu, and H.-S. Wong, "A Bayesian model for crowd escape behavior detection," *IEEE Trans. Circuits Syst. Video Technol.*, vol. 24, no. 1, pp. 85–98, 2014.

26. F. Baumann, J. Lao, A. Ehlers, and B. Rosenhahn, "Motion binary patterns for action recognition," in *Proceedings of 3rd International Conference on Pattern Recognition Application Methods*, pp. 385–392, Angers, France, 2014.

27. L. Kratz and K. Nishino, "Anomaly detection in extremely crowded scenes using spatio-temporal motion pattern models," in *IEEE International Conference on Computer Vision and Pattern Recognition*, pp. 1446–1453, Miami, FL, 2009.

28. S. Yi, H. Li, and X. Wang, "Understanding pedestrian behaviors from stationary crowd groups," in *Proceedings of IEEE International Conference on Computer Vision and Pattern Recognition*, pp. 3488–3496, Boston, MA, 2015.

29. S. Kumar, D. Datta, S. K. Singh, and A. K. Sangaiah, "An intelligent decision computing paradigm for crowd monitoring in the smart city," *J. Parallel Distrib. Comput.*, vol. 118, pt. 2, pp. 344–358, 2018.

30. P. Zhang, Y. Zhang, T. Thomas, and S. Emmanuel, "Moving people tracking with detection by latent semantic analysis for visual surveillance applications," *Multimed. Tools Appl.*, vol. 68, no. 3, pp. 991–1021, 2014.

31. A. Bansal and K. S. Venkatesh, "People counting in high density crowds from still images."arXiv preprint arXiv:1507.08445. 2015

32. J. Huang and W. Lee, "A smart camera network with SVM classifiers for crowd event recognition," in *Proceedings of the World Congress on Engineering* vol. I, pp. 13–18, London, UK, 2014.

33. G. Shu, "Human detection, tracking and segmentation in surveillance video," Dissertation, University of Central Florida, 2014.

34. P. Bilinski and F. Bremond, "Human violence recognition and detection in surveillance videos," in *13th IEEE International Conference on Advanced Video Signal-based Surveillance, (AVSS) 2016*, pp. 30–36, Colorado Springs, CO, 2016.

35. H. Wang, A. Kläser, C. Schmid, and C. L. Liu, "Dense trajectories and motion boundary descriptors for action recognition," *Int. J. Comput. Vis.*, vol. 103, no. 1, pp. 60–79, 2013.

36. D. McAllester, D. Ramanan, P. F. Felzenszwalb, and R. B. Girshick, "Object detection with discriminatively trained part-based models," *Computer (Long. Beach. Calif.)*, vol. 47, no. 2, pp. 6–7, 2008.

37. L. O'Gorman, Y. Yin, and T. K. Ho, "Motion feature filtering for event detection in crowded scenes," *Pattern Recognit. Lett.*, vol. 44, pp. 80–87, 2014.

38. R. V. H. M. Colque, C. Caetano, M. T. L. De Andrade, and W. R. Schwartz, "Histograms of optical flow orientation and magnitude and entropy to detect anomalous events in videos," *IEEE Trans. Circuits Syst. Video Technol.*, vol. 27, no. 3, pp. 673–682, 2017.

39. L. Zhang, Y. Feng, J. Han, and X. Zhen, "Realistic human action recognition: When deep learning meets VLAD," in *41st IEEE International Conference on Acoustic Speech Signal Processing*, pp. 1352–1356, Shanghai, China, 2016.

40. H. Rabiee, J. Haddadnia, and H. Mousavi, "Crowd behavior representation: An attribute-based approach," *Springerplus*, vol. 5, no. 1, p. 1179, 2016.

41. Y. Zhang, L. Qin, R. Ji, S. Zhao, Q. Huang, and J. Luo, "Exploring coherent motion patterns via structured trajectory learning for crowd mood modeling," *IEEE Trans. Circuits Syst. Video Technol.*, vol. 27, no. 3, pp. 635–648, 2017.

42. B. D. Eldridge and A. A. Maciejewski, "Using genetic algorithms to optimize social robot behavior for improved pedestrian flow," in *Proceedings of the IEEE International Conference on Systems, Man and Cybernetic*, vol. 1, no. 1, pp. 524–529, Waikoloa, HI, 2005.

43. M. S. Zitouni, H. Bhaskar, J. Dias, and M. E. Al-Mualla, "Advances and trends in visual crowd analysis: A systematic survey and evaluation of crowd modelling techniques," *Neurocomputing*, vol. 186, pp. 139–159, 2016.

44. Y. Yuan, J. Fang, and Q. Wang, "Online anomaly detection in crowd scenes via structure analysis," *IEEE Trans. Cybern.*, vol. 45, no. 3, pp. 562–575, 2015.

45. H. Singh, R. Arter, L. Dodd, P. Langston, E. Lester, and J. Drury, "Modelling subgroup behaviour in crowd dynamics DEM simulation," *Appl. Math. Model.*, vol. 33, no. 12, pp. 4408–4423, 2009.

46. Y. Xue, P. Liu, Y. Tao, and X. Tang, "Abnormal prediction of dense crowd videos by a purpose-driven lattice Boltzmann model," *Int. J. Appl. Math. Comput. Sci.*, vol. 27, no. 1, pp. 181–194, 2017.

47. S. Kumar, D. Datta, S. K. Singh, and A. K. Sangaiah, "An intelligent decision computing paradigm for crowd monitoring in the smart city," *J. Parallel Distrib. Comput.*, vol. 118, pp. 344–358, 2016.

48. J. Shao, K. Kang, C. C. Loy, and X. Wang, "Deeply learned attributes for crowded scene understanding," in *Proceedings of the IEEE Conference on Computer Vision and Pattern Recognition (CVPR)*, pp. 4657–4666, Boston, MA, 2015.

49. C. Zhang, H. Li, X. Wang, and X. Yang, "Cross-scene crowd counting via deep convolutional neural networks," in *Proceedings of IEEE Conference on Computer Vision and Pattern Recognition (CVPR)*, pp. 833–841, Boston, MA, 2015.

50. J. Wang and Z. Xu, "Spatio-temporal texture modelling for real-time crowd anomaly detection," *Comput. Vis. Image Underst.*, vol. 144, pp. 177–187, 2016.

51. M. Sabokrou, M. Fayyaz, M. Fathy, Z. Moayedd, and R. Klette, "Deep-anomaly: Fully convolutional neural network for fast anomaly detection in crowded scenes" [Online], 2016. Available http://arxiv.org/abs/1609.00866 [accessed 23 July 2018].

52. J. Wang, Y. Chen, S. Hao, X. Peng, and L. Hu, "Deep learning for sensor-based activity recognition: A survey," *Pattern Recognition Letters*, pp. 1–10, 2017.

53. A. Amato, V. Di Lecce, and V. Piuri, *Semantic analysis and understanding of human behavior in video streaming*, pp. 7–14, Springer: New York, 2013.

54. R. Agarwal, S. Kumar, and R. M. Hegde, "Algorithms for crowd surveillance using passive acoustic sensors over a multimodal sensor network," *IEEE Sens. J.*, vol. 15, no. 3, pp. 1920–1930, 2015.

55. L.-M. Ang and K. P. Seng, "Big sensor data applications in urban environments," *Big Data Res.*, vol. 4, pp. 1–12, 2016.

56. Y. L. Tan, V. Sehgal, and H. H. Shahri, "Sensoclean: Handling noisy and incomplete data in sensor networks using modeling," *Main*, pp. 1–18, 2005, Available https://www.sccs.swarthmore.edu/users/03/yeelin/docs/finalreport.pdf [accessed 17 January 2018].

57. H. M. Raafat et al., "Fog intelligence for real-time IoT sensor data analytics," *IEEE Access*, vol. 5, pp. 24062–24069, 2017.

58. R. Zhang, P. Ji, D. Mylaraswamy, M. Srivastava, and S. Zahedi, "Cooperative sensor anomaly detection using global information," *Tsinghua Sci. Technol.*, vol. 18, no. 3, pp. 209–219, 2013.

59. S. Mittal, "Online cleaning of wireless sensor data resulting in improved context extraction," *International Journal of Computer Applications*, vol. 60, no. 15, pp. 24–32, 2012.

60. M. Wisniewski, G. Demartini, A. Malatras, and P. Cudré-Mauroux, "NoizCrowd: A crowd-based data gathering and management system for noise level data," *Lect. Notes Comput. Sci. (including Subser. Lect. Notes Artif. Intell. Lect. Notes Bioinformatics)*, vol. 8093 LNCS, pp. 172–186, 2013.

61. B. T. Ong, K. Sugiura, and K. Zettsu, "Dynamically pre-trained deep recurrent neural networks using environmental monitoring data for predicting PM2.5," *Neural Comput. Appl.*, vol. 27, no. 6, pp. 1553–1566, 2016.

62. X. Ma, H. Yu, Y. Wang, and Y. Wang, "Large-scale transportation network congestion evolution prediction using deep learning theory," *PLoS One*, vol. 10, no. 3, pp. 1–17, 2015.

63. Y. Lv, Y. Duan, W. Kang, Z. Li, and F. Y. Wang, "Traffic flow prediction with big data: A deep learning approach," *IEEE Trans. Intell. Transp. Syst.*, vol. 16, no. 2, pp. 865–873, 2014.

64. X. Song, Q. Zhang, Y. Sekimoto, T. Horanont, S. Ueyama, and R. Shibasaki, "Intelligent system for human behavior analysis and reasoning following large-scale disasters," *IEEE Intell. Syst.*, vol. 28, no. 4, pp. 35–42, 2013.

65. Y. Zheng et al., "A cloud-based knowledge discovery system for monitoring fine-grained air quality," prepared Microsoft Tech Report, MSR-TR-2014-40, 2014 [Online]. Available http://research.microsoft.com/apps/pubs/default.aspx [accessed 23 July 2018].

66. L. Xia, R. Luo, B. Zhao, Y. Wang, and H. Yang, "An accurate and low-cost PM_{2.5} estimation method based on artificial neural network," in *20th Asia South Pacific Design Automation Conference (ASP-DAC)*, pp. 190–195, Tokyo, Japan, 2015.

67. R. Polikar, "Ensemble based systems in decision making," *Circuits Syst. Mag. IEEE*, vol. 6, no. 3, pp. 21–45, 2006.

68. R. K. Jain, J. M. F. Moura, and C. E. Kontokosta, "Big data + big cities: Graph signals of urban air pollution [Exploratory SP]," *IEEE Signal Process. Mag.*, vol. 31, no. 5, pp. 130–136, 2014.

69. A. N. Mohamed and M. M. Ali, "Human motion analysis, recognition and understanding in computer vision: A review," *J. Eng. Sci.* vol. 41, no. 5, pp. 1928–1946, 2013.

70. W. Raghupathi and V. Raghupathi, "Big data analytics in healthcare: Promise and potential," *Health Inf. Sci. Syst.*, vol. 2, no. 1–10, p. 3, 2014.

71. S. Amini, I. Gerostathopoulos, and C. Prehofer, "Big data analytics architecture for real-time traffic control," in *5th IEEE International Conference on Models and Technologies for Intelligent Transport Systems*, pp. 710–715, 2017.

72. L.-M. Ang, K. P. Seng, A. Zungeru, and G. Ijemaru, "Big sensor data systems for smart cities," *IEEE Internet Things J.*, vol. 4, no. 5, pp. 1259–1271, 2017.

73. E. Ahmed et al., "The role of big data analytics in Internet of things," *Comput. Networks Elsevier*, vol. 129, pp. 459–471, September 2017.

74. C. H. Lee et al., "Building a generic platform for big sensor data application," in *Proceedings of the 2013 IEEE International Conference on Big Data*, Silicon Valley, CA, pp. 94–102, 2013.

75. R. K. Ganti, F. Ye, and H. Lei, "Mobile crowdsensing: Current state and future challenges," *IEEE Commun. Mag.*, vol. 49, no. 11, pp. 32–39, 2011.

76. T. Franke, P. Lukowicz, and U. Blanke, "Smart crowds in smart cities: Real life, city scale deployments of a smartphone based participatory crowd management platform," *J. Internet Serv. Appl.*, vol. 6, no. 1–19, p. 27, 2015.

77. M. Q. Ngo, P. D. Haghighi, and F. Burstein, "A crowd monitoring framework using emotion analysis of social media for emergency management in mass gatherings," arXiv Prepr. arXiv1606.00751, December 2016.

78. P. M. Santos et al., "PortoLivingLab: An IoT-based sensing platform for smart cities," *IEEE Internet Things J.*, vol. 5, no. 2, pp. 523–532, 2018.

79. P. Bellavista, S. Chessa, L. Foschini, L. Gioia, and M. Girolami, "Human-enabled edge computing: Exploiting the crowd as a dynamic extension of mobile edge computing," *IEEE Commun. Mag.*, pp. 149–155, January 2018.

80. C. Martella, J. Li, C. Conrado, and A. Vermeeren, "On current crowd management practices and the need for increased situation awareness, prediction, and intervention," *Saf. Sci.*, vol. 91, pp. 381–393, 2017.

Chapter 4

Sustainable Crowdsensing Data Dissemination Using Public Vehicles

Rashmi Munjal, William Liu, Xue Jun Li, Jairo Gutierrez, and Marija Furdek

Contents

4.1 Introduction

In this chapter, we present the application of the mobile crowdsensing paradigm in supporting efficient, safe and green networking with a traditional network. We argue that today's mobile devices, with integrated or add-on sensors, can be efficiently used to crowdsource diverse information from different application domains. However, this proliferation of services is ultimately affecting our environment and leads to significant consumption of energy. Therefore, the reduction of energy consumption is becoming a major concern in wired and wireless networks because of the potential economic benefits and its expected environmental impact. These issues are usually referred to as "green networking," and they relate to embedding energy awareness in the architecture and product design, device usage and protocols of networks. The main concern is to maximize the performance of communication systems among the crowd while producing sufficient information with minimum energy costs. In order to address the challenge of energy efficiency, we propose a new data dissemination approach that features crowdsensing sensors (public transport), and we offload the crowdsensing data from wired and wireless networks to the scheduled public transport vehicles. The public transport vehicles are the moving sensors in this new communication paradigm, and they enable an alternative communication option in addition to the current infrastructure-based communications to reduce energy consumption and sustain quality of information (QoI). With the constantly increasing energy consumption for the transmission of growing amounts of data, more energy-efficient and sustainable solutions need to be incorporated into the future wired and wireless networks. Cisco [1] forecasted the mobile traffic rates as 49 exabytes per month for 2020, which means that the traffic will almost triple in the next 2 years. The total Internet traffic has experienced a dramatic growth in the past two decades, and it is still growing very fast. They also showed that the energy consumption of the information and communication technology (ICT) sector will be responsible for an enormous ratio (51%) of global energy consumption in 2030. The electricity consumption of communication technology is increasing and will be 30,715 TWh in the worst-case scenario by 2030. Therefore, its reduction is a critical and urgent issue for novel and innovative solutions.

Figure 4.1 shows the three digital laws: Kryder's law, Moore's law and Neilson's law [2], which state that new products come into the market with each passing year with new technology. The basic idea of Kryder's law is that technology advances double the storage capacity every 12 months. Moore's law is somewhat like Kryder's but refers to the processing speed of chips, which doubles every 18 months. Finally, Neilson's law estimates that bandwidth doubles every 21 months, so this last component of digital experience lags behind both storage and processing speeds. These three laws clearly explain that whatever new network technology comes into the market, the available data (in online storage) is never fully accessed by the new network technologies and the end users. There will always be a gap between the

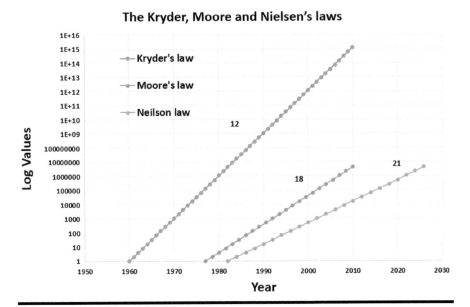

Figure 4.1 Kryder's, Moore's and Nielsen's laws.

available bandwidth and the available data or information storage online. This big data need will never be satisfied by the bandwidth of Internet technologies. Whenever there is a conflict between modern technology and users' desires, human desires always win, so the industry and researchers will need to keep advancing the set of Internet-related technologies. This nonstopping data needs lead to increasing levels of energy consumption, and thus more carbon emissions and harm done to the environment. To address this issue, we want to introduce a new option of utilizing the existing public transport networks. Our solution can pave a possible way out through decoupling the energy consumption from the data size and its transmission distance, even if that means that sometimes we need to sacrifice the delays associated with data delivery.

Many researchers have devoted their efforts to improving the resource efficiency of ICT systems, but most of them are still dedicated to study the telecommunications and Internet domains. Several attempts [3–5] in developing efficient, sustainable and integrated (wired and wireless) networks have been made. The concept of "opportunistic networks" (OppNets) is one of the techniques proposed to overcome this problem while disseminating crowdsensing data using a store-and-forward manner by connecting mobile devices. The key idea with OppNets is that the mobility of nodes may help to deliver messages, whether the network is connected or disconnected asynchronously in time [6]. Moreover, the Internet of Things (IoT) is developing from traditional homogeneous technologies with low resources to increasingly heterogeneous and resource-rich technologies.

Recently, there has been a fast-growing body of literature on energy efficiency and sustainable QoI levels while sending crowdsensing data; nevertheless, our work will focus on both perspectives simultaneously in a heterogeneous network. The proposed research focuses on crowdsensing data dissemination techniques, which will utilize existing public transport to improve the energy efficiency of data transport. Unfortunately, there is not much work related to the concept of energy efficiency in heterogeneous networks while offloading data from infrastructure-based parts of the network toward scheduled vehicles functioning as mobile nodes of an extended network.

We propose a novel model of crowdsensing data transmission in heterogeneous networks, where data will be offloaded from the wired and wireless networks toward the scheduled public transport networks, for example, buses, trams, trains, ferries and planes, by using delay-tolerant network (DTN) techniques. While transmitting such data from source to destination, people generally use two options: (1) a wired network, which forwards messages and gives guaranteed delivery of data, but while consuming more energy, and (2) a mobile or wireless network, where users use their mobile devices and forward messages using their devices, which also consume energy. We introduce a third option: to best utilize the existing public transport and vehicles and road infrastructure for sustainable communication and networking. We call these vehicles the local crowd. Thus, the communication and sharing of the crowdsensing data all happen in the local crowd, in order to reduce the data delivery delay and sustain energy usage.

4.2 Related Work

The research community's main aim is to maximize the performance of communication systems; however. on the other side, there is also a focus on reducing energy consumption [7]. Researchers are working on energy-aware communication technologies for heterogeneous networks with DTN-based approaches. Wireless communication networks [8] are expected to cooperate with wired networks for addressing performance challenges and energy utilization issues. Currently, the wireless communication medium works in a heterogeneous environment [9] with various wireless nodes (moving crowd and vehicles), and a wireless access network is defined as a wireless system that uses radio base stations (BSs) or access points (APs) to interface mobile terminals (MTs) with the core network or the Internet.

4.2.1 Mobile Crowdsensing

Mobile crowdsensing is a relatively new discipline, in which the users of modern smartphones use the rich sensing capabilities of their devices to collect and share information while on the move, as well as to form microcrowds around a certain crowdsensing activity.

With an increase in the number of mobile applications, the development of mobile crowdsensing systems has recently attracted significant attention from both academia and industry researchers. Mobile crowdsensing [10] is a technique where a large group of people (crowd) are involved; they share data, extract data and upload data onto the Internet through applications or directly via Internet connections. In other words, mobile crowdsensing goes through three different stages.

1. Data collection: The user collects data manually or automatically using wired or wireless networks. Some data has to be collected manually, and some other data gets collected automatically as soon as we open an application. In such a way, data sensing is triggered by a particular context, which will be predefined.
2. Data storage: The user stores that data within the device until it gets another user to share and communicate. For example, everyday traffic surveillance videos get stored within devices for a week, and then they will be shared with the council personnel, for example, to take action.
3. Data upload: The user uploads data using cellular or wired networks to share with others. Data can be uploaded automatically when the user comes into the range of wireless networks. It is under the control of users; they decide the uploading time, location and method to share. In our work, we will be using existing public vehicles, whose scheduled time is accessible to all users with predefined locations of bus movement. Therefore, we can better utilize them for sharing and communicating crowdsensing data.

Various research and development programs have been set up for achieving this magnificent challenge without compromising quality of service (QoS). This work is mainly focused on crowd communications while sustaining QoI attributes. However, due to the tremendously changing network topology, the crowdsensing data dissemination usually follows the strategy of "store, carry and forward." Moreover, the messages can only be opportunistically transmitted to their final destinations with some delay and smaller delivery probabilities. To improve the area of energy consumption and achieve good delivery ratios, researchers are concentrating on efficient message transmission techniques. This strategy is useful in the case of loose time requirements, but to sustain quality services, we will be using wired and wireless networks simultaneously. Yang et al. [11] explored a technique to predict the bus arrival time; they mainly observe the bus routes and record the sequences of observed cell tower IDs, which reduces the initial construction overhead. They do not require the absolute physical location reference. Wang et al. [12] analyzed the data delivery delay for the persistence of roadside unit (RSU) placement in a vehicular ad hoc network (VANET) through discontinuous connectivity. A mathematical model is established to describe the connection between the average delay for delivering road situation information and the placement distance between two neighbor RSUs.

4.2.2 Green Networking

Green networking is defined as the selection of energy-efficient networking technologies and better utilization of resources whenever possible. It has been estimated that 3% of the world's yearly electrical energy consumption and 2% of CO_2 emissions are caused by the ICT infrastructure [13]. Moreover, it is estimated that ICT energy consumption [14–16] is rising at 15%–20% per year. Specifically, 57% of the energy consumption of ICT businesses goes to users and network devices in mobile and remote networks [17]. There is a very broad area to validate the concept of green networking, which is defined in the following terms: (1) adaptive link rate, (2) interface proxying, (3) energy-aware infrastructures, and (4) energy-aware applications.

1. **Adaptive link rate:** As we know in the case of an Ethernet link, energy consumption is totally based on its utilization. The adaptive link rate has been designed in such a way that there will be low utilization of online resources and consumption will be less. This method can be used in two different ways: (a) switching links to low-energy-consumption mode or sleeping mode and (b) rate switching, which means the reduction of line rates, according to low utilization.

2. **Interface proxying:** The main aim of interface proxying [18] is to delegate traffic from the main board's CPU to low power consumption of the network interface card. Such data can be offloaded from one system to another to utilize the resources and reduce the consumption of energy.

3. **Energy-aware infrastructure:** This term can be further divided into energy-aware architecture and routing schemes [19]. In energy-aware architecture, a new approach suggests amending the structure used over an existing infrastructure. For example, the Grid5000 system is mentioned by many researchers; energy-aware routing aims to accumulate traffic over other network devices and allows other connected devices to be in an OFF mode as part of a rerouting strategy.

4. **Energy-aware applications:** Applications are divided into two categories: (a) user-level application and (b) kernel-level network stack. In user-level applications [20], a strategy is used to test peers whose status is not known. Peers present their energy state, and therefore we can judge whether it is energy-efficient or not. A peer does not wake up idle peers if it is not necessary. In kernel-level networks, many applications get shared by several peers. They store data and do not send it immediately, which helps to reduce energy consumption.

All over the world, various techniques are used to implement green networking. The Japanese government, for instance, has shown additional interest by introducing green IT for power efficiency, such as applications ranging from small-sized, low-scale observing to large-scale, energy-compelled environmental monitoring. In a network topology, energy is consumed at scanning, transmission, and data acquisition. All current approaches to green networking promise high energy savings at the cost of reducing the network performance. Since network performance indexes have been so far the only yardstick for operators and manufacturers, the proposal of such a trade-off may appear odd and hard to accept. Thus, the current challenge for most researchers involved in the area of green networks is to find specific solutions or mechanisms that could work with a negligible impact on the network-level performance.

The study reported in [21] classified various technologies to reduce energy consumption in wireless communications. New developments in smartphone technologies incorporate new architectures, new network management approaches, and novel forwarding techniques. Therefore, it is important to make a shift from traditional spectral efficiency toward energy-efficient designs and energy-preserving routing strategies. The future communications network is to be designed to offload traffic to other network devices while using comparatively fewer resources. Many researchers are working on energy efficiency techniques in wireless or wired infrastructure by proposing changes in hardware requirements or routing strategies. This work proposes a new solution with guaranteed delivery and using less consumption of energy in a heterogeneous network. One key component is based on the concept of a DTN with a capacity to store, carry and forward packets. We use DTNs where wired and wireless networks are not available or if there are looser time requirements.

4.2.3 Delay-Tolerant Network

DTNs [22] are a type of OppNet, which are being developed as a critical method to incorporate remote heterogeneous devices because of their convenience and adaptability of operation. Most of the devices used for communication have increased their storage capacities, processing and communication technologies; moreover, this advanced communication capacity has enabled a novel class of applications extending from mobile networks to vehicular networks [23]. These applications work on the principle of forwarding messages from one device to another and are known as OppNets. DTNs [24] are inadequate versatile networks, where an end-to-end path may not exist. The main principle is to store, carry and forward in the path toward the final destination. DTNs are often susceptible to the problem of high delay. Many application domains are taking advantage of the concept of DTN, for example, for crisis and disaster management, wildlife monitoring and transport engineering, to mention a few. Mostly, the traditional ad hoc networks and Internet routing protocols do not function admirably in the event of OppNets, where there is never a settled and dependable path between sender and receiver because of the node's mobility, network allotments, failure and other normal characteristics of a dynamic wireless environment [25]. The communication in a DTN depends on the contact opportunity between the nodes that emerge because of their versatility, and the store-convey and forward strategies. The message is gone to the intermediate nodes, which take it closer and nearer to the destination and, in the last hop, to the receiver itself. On the off chance that the intermediate node does not locate an appropriate location that guarantees taking the message nearer to the destination, it keeps the duplicate message in its buffer until it finds a reasonable node to pass messages to or finds the destination node itself [26]. Such types of networks result in long delays associated with the delivery of messages [23,27,28]. Recently, many applications have been emerging with the same concept of DTN. This concept is being advanced from the field of mobile ad hoc networks (MANETs), but the basic difference between DTN and MANET nodes is that the latter use TCP/IP for communication and DTNs use an application-layer bundle (transport and network layer), and they also use the concept of store-carry and forward with other peers. It may take a long time to find an appropriate relay node, so there is a need to use message carriers to store messages in the buffer for a long time. Many opportunistic routing algorithms [26] have been defined in terms of forwarding messages to the destination.

4.2.3.1 Routing Protocols Used for DTNs

- **Epidemic routing:** This is a forwarding protocol [29] that works on the concept of spreading a message like a disease. A node starts spreading the message to all its neighboring nodes and keeps on repeating it until the message reaches its destination. It consumes huge resources and provides optimal routing performance in terms of delivery ratio while minimizing delay. In this algorithm, the author considered a case of 45 nodes using the random way point model and their traffic patterns. They placed an upper bound on message hop count and buffer space according to a single node. In this way, incrementing bounds on all these parameters and applications just increased the probability that the message would be successfully delivered, even if they had to consume more resources. The advantage of this method is its robustness: random pair-wise message dissemination among all nodes ensures message delivery. This routing protocol involves two steps:
  - Exchange of summary vectors: Every node contains an index of messages stored in its buffer, which include unique message IDs.
  - Exchange of messages: One node computes a set of messages carried by other nodes; the same is done by other nodes as well. If such a computed node is not empty, it requests its peer to transmit a message with its corresponding IDs.
- **Direct delivery:** In this protocol [30], the source node does not pass the message to potential forwarding nodes; rather, it keep it within itself until it comes directly in contact with the destination node. This plan is straightforward, simple to send and uses the least data transmission and system assets for message exchange, as every message is transmitted at most once to the

destination node. Then again, there may be a long deferral for message conveyance when the source never meets the destination, or there may not be a quick contact between the source and the destination; however, a path exists through the center of the road nodes for message passing. On the off chance that the source hub comes up short, then the message will be lost due to a duplicate message in the system. In this plan, the likelihood of conveyance is poor and not adequate for circumstances where high conveyance likelihood is needed.

■ **Spray and wait:** This algorithm [31] is divided into two parts: (a) spray phase and (b) wait phase. In the spray phase, many message copies are generated from the source and spread toward distinct nodes. They use direct transmission toward the destination. In the wait phase, if the destination is not discovered, each of the distinct nodes that received a copy uses the concept of direct transmission to reach the final destination. Initially, this phase uses the concept of epidemic routing, but it is much better than the epidemic method in terms of generating low contention, specifically when a high traffic load flows. This system is also highly scalable. This strategy outperforms all existing algorithms in terms of transmission and delivery delays.

■ **Single-copy algorithm:** Spyropoulos et al. defined single-copy-based protocols for an OppNet where nodes forward a message only if they know the destination and transmit a single copy of the message once. Their [32] utility function defined the usability of nodes according to the number of meetings among all nodes and taking into account their last encountered information. As soon as a high-utility node is revealed, they use the approach of seeking and finally reach the destination. Nodes make local forwarding decisions based on their connectivity and the prediction of future connectivity details. A combination of a simple random policy and a utility-based policy is followed to reach the final destination.

■ **PRoPHET:** PRoPHET (Probabilistic Routing Protocol using History of Encounters and Transitivity) [33] uses nonrandom mobility and contact patterns with a probabilistic metric called delivery probability. It considers the history of previous contacts, which are not to be considered random, to identify mobility patterns. This model is based on the prediction of the probability of each node for all known destinations. The predictability of delivery is denoted as $P_{A,B}$, and its range is defined as $0 \leq P_{A,B} \leq 1$.

4.2.4 Mobile Data Offloading to Local Crowd (Public Vehicles)

Vehicular DTN (VDTN) [34] is based on the concept of DTN. It handles nonreal-time applications at low cost and under unreliable conditions. Vehicles can be used as a data carrier between terminal nodes either in rural areas or in emergency scenarios. VDTN follows the Open System Interconnection reference model. This protocol works on two layers of the OSI model, the physical layer and the data link layer, for functioning and data management. A data plane aggregates data into bundles and transports it to the destination. At the source node, a bundle aggregation layer aggregates packets and then destination addresses get converted into VDTN terminal node addresses. Many techniques can be used for data assembly. Therefore, this strategy can help in vehicle-to-vehicle communication.

We subcategorize VDTN into (1) using scheduled vehicles (i.e., public transport) or (2) using local crowd facilities. Many researchers have worked on the scheduled vehicle option to make networking more efficient as that category avoids random movement of the vehicle. Our research also employs public transport plus the traditional way of communicating in order to take advantage of predefined schedules and routes for particular locations. Crowdsensing data can be offloaded onto a public transport vehicle to use it as a data carrier to transport the packets to another location.

Gorcitz et al. [35] used vehicle carriers for big data transfer. Due to the increased Internet traffic in recent years, they decided to exploit the existing worldwide road infrastructure as an offloading channel to help the legacy Internet assuage its traffic loads. Motivated by the need for technical flexibility and cost-effective scalability, large companies, organizations, universities and governmental agencies constantly move their data and applications within and between data centers to balance workloads, handle replication and consolidate resources. They put their efforts into using conventional vehicles for their proposed offloading scheme, which generates cost reductions and improves capacity. Assisted-DTN architectures involve various data carriers or forwarders, ranging from buses to airplanes, to compensate for the lack of continuous connectivity by bridging disconnected nodes. They justify their proposal by saying that vehicles help to reduce traffic and costs while delivering data within reasonable time frames.

While moving further with the concept of offloading data, Cheng et al. [36] discussed vehicular Wi-Fi offloading. Vehicular offloading relies on drive-thru Internet access opportunities; drive-thru Internet access [37] means the Internet provided by all roadside-placed APs to the moving vehicles. This Internet access works using threefold policies: First, a highly mobile vehicle gets connected quickly, which ultimately limits the transfer of data in one connection. Therefore, the probability of packet losses is higher. This offloading scheme helps to sustain a connection for a long time for nonvehicle users. In the second fold, vehicle users meet many APs with different connection times, and finally, the offloading scheme adds in the prediction of Wi-Fi availability to transfer data. To increase the efficiency of drive-thru Internet access, many more modifications have been proposed in the literature. In [38], a new mechanism was introduced to improve the connection time, which enhanced the data transfer's performance. They merged all the processes into one process and have a "name the connection" establishment phase, where a network-independent identifier is used by both host and vehicle users. Moreover, this mechanism keeps sending probe packets periodically to distinguish wireless losses from congestion losses. In 2010, the authors introduced [39] an offloading scheme named "Wiffler," which determines whether to defer applications onto the Wi-Fi network rather than cellular networks. This strategy introduced delayed offloading when both 3G and Wi-Fi networks were available. This switching is based on the user's preference and information port names. Wi-Fi throughput varies according to delay periods and the predictions made according to the estimated number of APs. Whitbeck et al. [40] proposed an infrastructure offloading scheme named push-and-track, which determines how many copies should be disseminated, to which devices the copies should be sent and at what times the copies should be disseminated. When nodes get content from infrastructure-based nodes to mobile devices, that content is sent in an epidemic manner with acknowledgment sent back to the infrastructure network. The time to live (TTL) value is used to determine the number of copies to be pushed from the infrastructure nodes.

In [41], the authors used a new routing technique to offload data from an infrastructure-based network toward a DTN. Initially, the messages start disseminating from the infrastructure-less network, and as the probability of successful delivery decreases, the process shifts to the infrastructure-based network to ensure the delivery of the data. These decisions are based on many factors, like a fraction of the infrastructure-capable devices and message size. Unicast end-to-end communication is done between all mobile devices with the hybrid routing system. Rebecchi et al. [42] introduced many offloading techniques for cellular networks to reduce network congestion. These offloading techniques are categorized as delayed offloading and nondelayed offloading.

1. Nondelayed offloading: Nondelayed offloading provides obvious solutions, such as the use of Wi-Fi hotspots to lessen congestion and cost. Cellular BSs and terminal-to-terminal (T2T) approaches use wireless technologies to establish direct communication between devices and cellular-to-device communication. AP-based offloading is a user-driven approach.
2. Delayed offloading: In delayed offloading, content reception can be delayed up to a certain point of time, but with good delivery conditions.

Now, all mobile phones and laptops are increasing data traffic every year, and this trend is expected to continue further with all new developments. Therefore, Dimatteo et al. [43] propose a solution to offload traffic from cellular networks toward metropolitan Wi-Fi APs, which is known as the MADNet architecture. While transferring bulk data, this architecture is really beneficial even in the case of sparse Wi-Fi networks. MADNet consists of a total of six modules: a connectivity module, a location module, a protocol module, naming and forwarding modules, and a data module. All these modules are responsible for providing connectivity to users according to location, name and then forwarding data. Every node of the system generates data to upload and requests content to be downloaded on another side. In such a way, at the time of downloading content, a user request to the nearby BS with all the details (position, speed and direction), which is known for all app positions. The BS makes a list of all APs in that route and nodes consult the particular AP to download the content. This list keeps a record in storage for future direct communication with different interfaces such as Bluetooth/Wi-Fi (Table 4.1).

Here, we introduce a scheduled vehicle as a part of the communication if there are flexible delivery time requirements. Moreover, there are many rural areas without a wired or wireless network infrastructure. Then, these scheduled vehicles can help to deliver messages with some delay. However, we cannot ignore our traditional way of communication. In some cases, our messages can be lost when using a DTN approach, and we do not want to compromise when a high QoS is required. Therefore, we will use a heterogeneous network combination to make communications more effective and energy-efficient.

4.2.5 Heterogeneous Network

Several wireless communication technologies have been developed so far. Bejerano [44] presented work on efficient and low-cost infrastructure for connecting multihop wireless and wired networks. On these networks, some nodes behave like an AP and function as a gateway between the wired and wireless backbones. This work is subdivided into two phases; one is to design optimization where nodes are further partitioned into clusters and they select a single AP node at each cluster. The operational part includes messages to be delivered through a spanning tree rooted at the AP of each cluster, which is called an adaptive delivery mechanism.

Lei and Perkins [45] proposed an integrated network (ad hoc network and Internet) with the use of the routing information protocol (RIP). It routes and forwards the datagram to a particular destination appropriately, but without keeping track of a return path. Therefore, QoS constraints are not satisfied. Jönsson et al. [46] introduced an architecture called MIPMANET (Mobile IP for MANET), which uses a MIP (Mobile IP) foreign agent and the AODV (ad hoc on-demand distance vector) protocol. Tseng et al. [47] proposed an architecture that uses standard IP routing to relay IP messages and data packets without using MANET routing. All information exchange gets done by delivering Mobile IP messages without exchanging routing tables or Address Resolution Protocol (ARP) messages. The main disadvantage of the protocol is the cost of flooding advertisements, which makes it nonsustainable for infrastructure networks. Hasan et al. [13] used a stack-based approach and a topology-based approach. In the stack-based approach, the level of integration between the Internet and the wireless sensor network depends on the similarities between their network stacks. On the other hand, in the topology-based approach, the level of integration depends on the actual location of the nodes that provides access to the Internet. From this, many solutions have been taken into account to integrate both networks, such as TCP/IP, SCADA (supervisory control and data acquisition) and first responder systems. All such solutions

Table 4.1 Overview of Offloading Schemes

Work	Message Direction	Goal	Offloading Type	Mobile Device Types
MADNet [43]	Infrastructure → mobile device	Reduces infrastructure load	Delayed	All infrastructure access, all DTN
Drive-thru Internet [37]	Infrastructure → vehicle	Reduces infrastructure load and stronger connection	Nondelayed	All vehicle, Wi-Fi access
Wiffler [39]	Infrastructure → mobile device	Prefers Wi-Fi over cellular	Delayed	All Wi-Fi and cellular capable
HRS [41]	Mobile device ↔ mobile device	Offloads infrastructure, prefers DTN	Delayed	Heterogeneous infrastructure access/ ad hoc
Kashihara [42]	Wireless nodes → vehicles	Reduces traffic congestion	Delayed	All Wi-Fi and cellular capable
Push-and-Track [40]	Infrastructure → mobile device	Reduces infrastructure load	Delayed	All infrastructure access, all DTN

are applicable according to different substations and different situations. The major drawback is the security issues while integrating the wired and wireless networks.

In [48], the researchers focus only on the inefficient handling of delay-tolerant traffic. They want to shift traffic from peak hours, taking advantage of the delay-tolerant nature of some types of traffic. Therefore, two main strategies are used: (1) offering incentives under flat rates to those end users who shift their delay-tolerant traffic (this is accomplished by providing a higher than purchased access rate as a bonus) and (2) augmenting the network with additional storage, such as an Internet post office.

HYMAD: The authors proposed a HYMAD (Hybrid DTN-MANET routing) [49] protocol for dense and highly dynamic wireless networks, which uses the concept of DTNs between a disconnected group of nodes while using MANET routing strategies. The best part of this strategy is its capacity to make connections according to the connectivity patterns of the network. Intragroup delivery is done by using an ad hoc routing protocol, and the DTN protocol contributes to intergroup delivery. According to the connectivity characteristics, the system decides its behavior to make a connection. In the case of dense connectivity, it functions like a traditional MANET. When connectivity is very sparse, it behaves like a traditional DTN routing protocol. The motivation is to get good delivery in a heterogeneous network.

HSBR: Enhanced hybrid social-based routing [50] is an optimal solution for combining features of both dynamic source routing (DSR) [51] and social-based opportunistic routing [52,53]. This strategy is used to achieve a high success of packet delivery between the source and destination from different start conditions and different paths of mobile networks. A DSR path is used to request a route and for opportunistic sending of packets. Data transmission follows the social relationship between nodes. In this work, a modified version of DSR is used that follows a route discovery or request, route maintenance and detection of the relay node. Social-based opportunistic routing is based on the relationship between nodes according to their personal or private profile and contact profile (address book). HSBR establishes delivery of the message when the standard MANET routing does not work properly or in a disaster situation. Another of its advantages is to deliver messages with the expectation of delay. In [54], the authors focused on hybrid networks that are based on overlay and DTNs. DTNs integrate with the partial infrastructure to cooperate in transmissions. Infrastructure devices work as a proxy for other devices to take messages through an overlay. Communication can be done with any type of infrastructure-capable device (temporal infrastructure or continuous infrastructure access).

The IoT [55] is changing continuously according to varying demands and requirements; therefore, this work expanded existing routing algorithms with a new routing scheme, which consists of a heterogeneous set of nodes while taking into account two parameters: (1) delivery capability (L) and (2) number of copies (C). This work focuses on the use of opportunistic routing to provide various services, such as traffic control, security, environmental issues (in terms of CO_2 emission) and making enterprise processes more efficient and effective. Performance is much better with the overall use of RSUs as high delivery nodes to service a large number of copies to all pedestrians. The given parameters help us to do comparisons in terms of delivery ratio and latency with an improved version for various types of sensor nodes. The expected delay in the optimal case, when $L = \infty$, is

$$ED_{opt} = \frac{H_{M-1}}{M-1} ED_{dt}, \text{ where } H_n = \sum_{i=1}^{n} \frac{1}{i} \text{ and}$$

M is the number of sensor nodes. Many opportunistic routing [56] protocols have been proposed, but all of them get differentiated in terms of assumptions and the types of the networks used for their evaluation. A heterogeneous architecture is defined, including fixed infrastructure, mobile infrastructure and mobile nodes.

1. Fixed infrastructure: Here, RSUs get positioned along main roads of a particular area. These are fixed with lampposts, Global System for Mobile communications (GSM) BSs and walls. They work as a backbone network that connects mobile nodes like phones with central servers, which distributes information from central servers to other regions. Short-range Bluetooth and Wi-Fi 802.11 are the interfaces used for communication. Bluetooth is used in the case of the RSU and regular phones for low range. The Wi-Fi interface helps to communicate between buses, trams, cars and smartphones.
2. Mobile infrastructure: Buses and trams with predefined route information come under this category. Buses and trams move fast; therefore, Bluetooth is not the best mode for them.
3. Mobile nodes: Mobile nodes are those components that consist of cars and mobile phones carried by walking pedestrians. Here, we cannot predict a path or route because of their unpredictable movement. These are classified into smartphones or regular phones that use Wi-Fi and Bluetooth interfaces.

In [57], the authors discussed metropolitan environments with the intention to provide delay-tolerant services to areas where end-to-end connectivity is not possible. It utilizes the CARPOOL plan to connect between ferries and gateways to compute routes to online gateways. Therefore, free public Internet access Wi-Fi hotspots were deployed through public transport, such as ferries, buses and trams. DTN gateways are located in an offline mode near all ferry stops to get an Internet request from all end users and in such a way as to act as a relay node. With prior knowledge of contacts between gateways, a high delivery ratio with minimum overhead was achieved. Therefore, their primary focus is on the scheduling of the ferry to access the Internet, the schedule of ferries and the energy-efficient design of all computing operations. This method, however, does not adapt well to expansive deviations from the predefined plan and is not sufficiently adaptable to exploit opportunistic contacts between ferries.

In 2015, Komnios and Kalogeiton [58] enhanced their work as a smart mechanism of a routing and connectivity plan with the name CARPOOL+. This plan is being introduced for urban transport with prior knowledge of contacts between gateways and ferries to compute routes with traffic. It is a simple-to-convey design that exploits both prescheduled contacts in public transport networks and opportunistic contacts among ferries to provide delay-tolerant Internet access to end users. In such a way, it provides a smart mechanism to select a route for the earliest delivery time with less replica overhead ratio. It mainly concentrates on two dynamic mechanisms: (1) opportunistic contacts between ferries and providing a better route for earliest delivery, and (2) path recalculation if there is any change due to traffic congestion. Route selection depends on several aspects:

1. As soon as end users make Internet requests to the offline gateway, CARPOOL+ selects an online gateway and classifies the best suitable gateway for this bundle; even when a middle gateway receives a bundle, it recalculates the path and updates its header to proceed further.
2. Even if a ferry is later than the scheduled time, it downloads bundles predestined to the particular gateway and recalculates the path.
3. In the meantime, offline gateways make the decision to recalculate the path prior to the next scheduled ferry to forward messages according to current conditions or opportunities if any ferry is delayed.
4. Even if there are two ferries in range, the decision will be made according to the earliest delivery time available for the bundles to forward.

In [59], the authors provide a promising solution called cost-effective multi-mode offloading (CEMMO), with offloading of data to the best possible choice among the following three options to reduce overall costs in terms of energy efficiency, financial settlement and user satisfaction:

1. Cellular delivery (3G network): They assumed that the 3G network is always available and that success probability is high in this option.
2. Delay-tolerant offloading: Here, every user's probability to get Wi-Fi to upload data before the delay-tolerant indicator (DTI) expires.
3. Peer-assisted offloading: Data will be offloaded through intermediate mobile devices. It is used not only for offloading data but also for its content and popularity.

In such a way, while applying the mobility and connectivity plan of peers based on a Markov process, CEMMO gives authority to the cellular operator to make the best decision about the most effective mode of communication according to cost minimization. In such a way, the offloading percentage is 59% toward mobile data traffic and the cost reduction calculation is 16% over delay-tolerant offloading (DTO), which is an improvement of 31% on energy consumption. In this model, user mobility during any time interval *t* works on the concept of the mobility prediction model; the next region that will be visited by a user during a time interval is assumed to solely depend on the previous one for each user.

- $N(X, t)$ = The number of visits in region X during time interval t (i.e., starts at time t)
- $N(A \rightarrow X, t)$ = The number of transitions from A to a neighboring region X during t
- $_{duration}$ = The duration of each time interval
- $t_{ext} = t_{duration} + t$ as the start time of the next time interval.

The probability of a user located in region A moving to region X directly within t is

$$P(X|A,t) = N(A \rightarrow X,t)/N(A,t)$$

The probability that the user stays in region X until the end of time interval t is

$$P_{stay}(X,t_{next}) = P(X|X,t) = N(X \rightarrow X,t)/N(X,t)$$

Therefore, a cost-effective decision can be made according to the offloading ratio, the average cost per megabyte and the average cache size. The offloading ratio expresses the fraction of the total amount of generated data that is offloaded through Wi-Fi networks. The average cost per megabyte is calculated as the sum of the total data (in MB) that was transferred through each transfer policy multiplied by the cost of each policy, divided by the total data transfer:

$$\text{Avg Cost} = (\propto \times \text{Dat Transmitted 3G} + \beta \text{ Data Transmitted Wifi})1$$
$$/(\text{Total Data Transmitted})$$

where α is the transferred cost per megabyte through 3G and β is the transfer cost per megabyte through Wi-Fi.

The proposed work focuses on an energy-aware data dissemination technique by utilizing existing public transport. To encourage users to use our technique, we will have to design an incentive mechanism for them. Incentive mechanisms basically attempt to offer a benefit that outweighs the cost for each network participant. Many researchers have already used incentive mechanisms to promote users' participation. These incentives are usually divided into two parts: (1) monetary incentives and (2) nonmonetary incentives.

4.3 Crowdsensing Data Dissemination Decision among Three Layers

In this chapter, we introduce a sustainable crowdsensing data dissemination technique that leverages existing public transport networks to extend the overall Internet connectivity while reducing energy consumption with acceptable data transmission delays. The Internet traffic congestion problem, energy consumption, poor networks and many more problems are being faced by users while using traditional networks. However, users generally have the first two layers, such as wired and wireless networks, at their disposal as an available technology to disseminate data. This third layer will help to sustain QoI and make communications better by the moving crowd and by sharing their data utilizing public transport (Figure 4.2).

Some of the available technologies are wired networks, wireless networks, DTNs, overlay and scheduled vehicle or public transport. We utilize all the available technologies to make an optimal decision. Our research model adopts the following principles.

Principle: Before embarking on any architectural design, it is useful to identify the principles embodied in the architecture. These principles allow us to intelligently navigate the infinite space of possible designs. We believe that these principles are applicable to any realistic architecture that uses mechanical backhaul and has goals substantially similar to ours. Our contribution is to take the parameter index as useful parameters to consider and make an optimal data dissemination decision among the three options according to performance metrics.

1. **Energy-efficient crowd interaction among three layers:** Energy-efficient transmission is possible if we utilize all resources efficiently for crowd interaction. Wired and wireless networks communicate through the core network

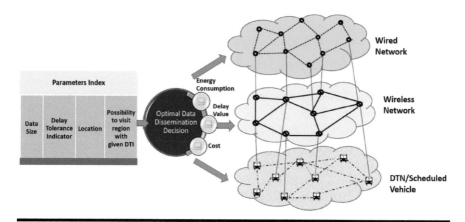

Figure 4.2 Optimal data dissemination decision according to performance metrics among three layers.

with a cellular link and Wi-Fi link. The third option is using public vehicles to carry data from one location to another. Vehicles are not emitting any extra energy if they are being used as a data carrier and can be utilized efficiently to save energy consumption.

2. **Cost through sharing core infrastructure:** Cost is also an important factor to consider; we have three options to select from when sending data, and we compare the options according to the users' demands and their cost-effectiveness characteristics when taking into consideration all the defined parameters. We believe that our new third option will definitely achieve the target in terms of cost as well.

3. **Decoupling:** In both the economic and environmental fields, decoupling is becoming an increasingly used concept in the context of economic production and environmental quality. It refers to the ability of an economy to grow without corresponding increases in environmental pressure. In our model, decoupling demonstrates a significant approach in terms of energy consumption and data volume. Here, it refers to the relative growth rates of carbon emissions generated by telecommunication and Internet networks, their impact on the environment and the relevant variables of data volume and distance.

4. **Utilizing existing public transport as a sensor for crowdsensing:** We are utilizing existing public transport as a sensor for crowdsensing to make communication more energy-efficient and effective. There is neither extra cost for arranging new vehicles nor extra carbon emissions to carry data. The vehicle may have some storage room to store data, and it may involve a one-time investment in the beginning, which will be compensated when used over several years.

5. **Acceptable delay:** DTN topologies can be used in the case of rural areas and where other networks are not available. Moreover, if data is delay tolerant, the third layer can be fully utilized to make network communications more energy-efficient. In some cases, this option could introduce less delay than the core network.

4.3.1 Case Scenario 1

We introduce a new layer to disseminate crowdsensing data with DTN features, that is, networks that work on the paradigm of store-carry-forward. We consider scheduling buses to carry data from one place to another. The user will offload delay-tolerant data to a scheduled bus or shuttle to deliver it to the destination. For example, Auckland University of Technology (AUT, our university) sends everyday large amounts of data from the AUT city campus to the AUT south campus. This data includes backup copies, information sharing and content delivery, and it is sent through the organization's network infrastructure (wired or wireless network). AUT has another option to best utilize its existing transport infrastructure: AUT shuttles. An AUT shuttle bus goes every 30 minutes from the city campus to the south campus (and returns) and can be utilized as a part of data communications in a more energy-efficient manner, as shown in (Figure 4.3). There are two possible ways to send data from the city to the south campus, which are discussed in the following:

- **Wired or wireless network (Option A):** In this option, AUT can upload their data to the network through a wired or a wireless connection, and on the other side, it will be downloaded from the network. Energy consumption in this option depends on energy consumed in downloading and uploading on both sides and the number of hops used to transmit and forward messages to the final destination. It can be calculated using Equations 4.1 and 4.2 and is considered 0.2 kWh/GB [60]. The cost factor is the financial cost involved, according to power consumption, and it can be considered from Equation 4.5. The delay value depends on the bandwidth used and storage capacity, which can also be examined with Equation 4.3.
- **Scheduled bus or AUT shuttle (Option B):** Option B is another option of crowdsensing data dissemination; it is delay tolerant and can be best utilized to communicate by offloading data onto an AUT shuttle to save energy. The AUT shuttle buses can have one storage cabinet installed onto them, and data can be offloaded from the network infrastructure toward the road infrastructure. As we are utilizing the existing road infrastructure, there is no extra cost to arrange a vehicle to carry data to the destination. Energy consumption in this option depends on energy used while uploading data to the vehicle, energy used in travel and energy consumed in downloading at the destination, as defined in Equation 4.2. There will be a one-time investment for the storage cabinet, but no extra cost for transmission, as defined in Equation 4.6. Delay is calculated in Equation 4.4 and is defined as the delay in offloading data to the vehicle with Wi-Fi Direct (64 MBps), the delay in traveling and the delay in downloading data from the vehicle to the destination.
- **Sensitivity analysis:** We carried out a sensitivity analysis that covers the traditional way of communications and our solution (scheduled vehicle) to know the contribution of third layer. In our case scenario, both locations are 20 km apart. The following performance metrics will be considered, with energy intensity values defined in Table 4.2 [61].

4.3.1.1 Performance Metrics

1. **Cost:** Cost is defined as the financial cost involved in message transmission.
2. **Delay:** Delay is the delay encountered by a packet when it is delivered to the final destination.
3. **Energy consumption:** Energy consumed is calculated as the amount of energy spent on transmission and scanning.

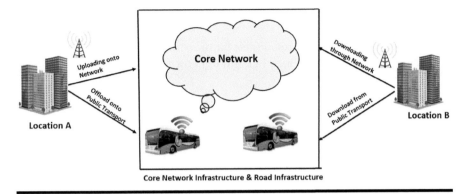

Figure 4.3 Crowdsensing data dissemination through AUT network and road infrastructure.

Table 4.2 Energy Intensity of Core, Access, 4G, Wi-Fi and Wi-Fi Direct

Network	Energy Intensity
Access network	0.2 kWh/GB
Core network	0.019 kWh/GB
4G mobile	4.65 J/MB
Wi-Fi	2.85 J/MB (at a distance of 1 m)
Wi-Fi Direct	2.85 J/MB (at a distance of 1 m)

$$E_{\text{total}(a)} = E_{Co} * V_{\text{data}} \tag{4.1}$$

$$E_{\text{total}(b)} = \left(E_{\text{wifi-di}} + E_{\text{shuttle}} + E_{\text{wifi-di}} \right) * V_{\text{data}} \tag{4.2}$$

In Equations 4.1 and 4.2, energy consumption for both options is calculated, where

E_{co} = Energy consumed in the core network
$E_{\text{wifi-di}}$ = Energy consumed in offloading data onto the shuttle with Wi-Fi Direct as an interface
E_{shuttle} = Energy consumed by the shuttle while traveling
V_{data} = Volume of data

$$D_{\text{total}(a)} = \frac{V_{\text{data}}}{B_{co}} \tag{4.3}$$

$$D_{\text{total}(b)} = \frac{V_{\text{data}}}{B_{\text{wifi-di}}} + \frac{V_{\text{data}}}{B_{co}} + \frac{V_{\text{data}}}{B_{\text{wifi-di}}} \tag{4.4}$$

In Equations 4.3 and 4.4, the delay is calculated for both options, where

B_{co} = Bandwidth value responsible for the core network
$B_{\text{wifi-di}}$ = Bandwidth value used in the Wi-Fi Direct link using the Wi-Fi Direct interface
V_{data} = Volume of data

$$C_{\text{total}(a)} = C_{Co} * V_{\text{data}} \tag{4.5}$$

$$C_{\text{total}(b)} = (C_{\text{wifi-di}} + C_{\text{shuttle}} + C_{\text{wifi-di}}) * V_{\text{data}} + C_{\text{investemen}} \tag{4.6}$$

In Equations 4.5 and 4.6, the cost involved is calculated for both options, where

C_{Co} = Cost of the core network
$C_{\text{wifi-di}}$ = Cost involved in offloading data onto the shuttle with Wi-Fi Direct as an interface
C_{shuttle} = Minimal cost consumed by the shuttle while traveling
V_{data} = Volume of data
$C_{\text{investment}}$ = One-time investment cost to install storage room on the bus

Figure 4.4 shows that energy consumption is much lower when using the scheduled vehicles than with the wired or wireless network. It keeps on increasing per data volume, but our solution consumes much less than other options. Figure 4.5 gives a detailed description of the financial costs involved in communication in

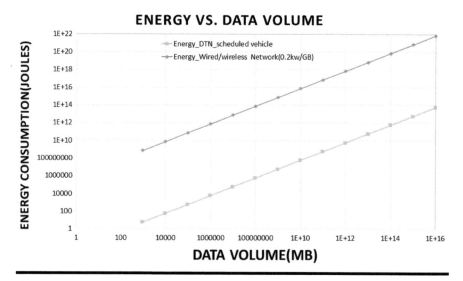

Figure 4.4 **Energy consumption in wired, wireless and scheduled vehicle networks.**

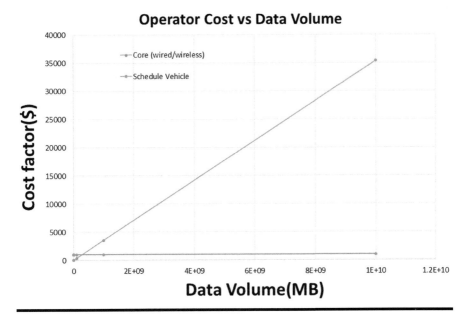

Figure 4.5 Cost factor included in wired, wireless and scheduled vehicle networks.

each of these three layers, and shows that the cost of the scheduled vehicle scenario is lower than that of the wired and wireless networks, with a one-time investment for the storage cabinets in the shuttle buses.

Figure 4.6 shows the delay values with respect to the data volumes. Delay values have some threshold points to prove that they will be acceptable as data volumes increase, and they are less than those of other networks. Finally, all the performance metrics satisfy the condition that scheduled vehicles (public transport) can be a better solution in heterogeneous networks if data is delay tolerant.

4.3.2 Case Scenario 2

To elaborate, case study 2 talks about an individual user working from the AUT city or south campus. We consider the same AUT shuttle as a third layer with the traditional way of communication. Here, the user has two traditional options to send crowdsensing data, either via a wired network with a 4G cellular link or using a Wi-Fi link to connect to their device. In Figure 4.10, John intends to send data to Bobby from AUT's city campus to AUT's south campus, the distance between both places is 20 km. He has three options to send the data.

- **Wired network (Option A):** In this option, John uses the fixed or core network infrastructure with the 4G access network for communication with a cellular link between a device and the network. He makes a zip folder and

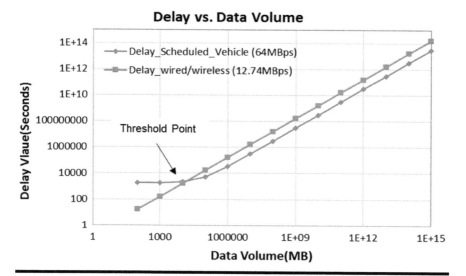

Figure 4.6 Delay value included in wired, wireless and scheduled vehicle networks.

sends crowdsensing data over the Internet or just attaches the file and sends an email from his location. This data is to be transmitted from his mobile device (e.g., smartphone or laptop) to the nearby mobile BS, which then transmits it to the core network. On the other hand, Bobby is to receive the data through the cellular BS near him. This is the traditional way of sending data from a source to the final destination. Energy consumption, in this case, is according to the energy intensity values for the core network and includes the usage of a cellular link. Delay values also depend on the data rate and data volume and can be calculated from Equation 4.10. Cost factors can also be calculated by Equation 4.13.

- **Wireless network (Option B):** Many wireless networking topologies are available to share data between two users. For example, users follow point-to-point communication/Bluetooth or Wi-Fi. Both of the users take advantage of AUT's free Wi-Fi for accessing the network. We assume the same or similar core network connectivity between the two APs with a Wi-Fi link. Energy, delay and cost values can be calculated from Equations 4.8, 4.11 and 4.14, which depend on data volume and energy intensity values.

- **Scheduled vehicle or public transport (Option C):** This is the option that we are proposing: a new way of data transmission by using delay-tolerant techniques to offload traffic from the core network to the transport network. John can utilize the existing AUT shuttle to carry his data to Bobby. There is a regular shuttle bus running between the AUT city and south campus, and the frequency is about every 30 minutes in both directions and the travel duration is about 30 minutes most the time during the daytime. Our scheme tries to take advantage of this existing transport infrastructure. Between mobile devices, Wi-Fi Direct is considered the communication mode to offload or download data onto the bus. Moreover, John's data will transfer to the shuttle first, and then the shuttle will carry the data to the destination, by assuming that this data has a delay-tolerant nature; for example, it can tolerate a delay of at least 2–3 hours. In this option, the energy consumed will be in offloading and downloading the data from the device to the bus and from the bus to the device at the destination. It can be calculated by Equation 4.3. The delay and cost values also depend on the data rates and the one-time investment of a storage cabinet on the bus, as defined in Equations 4.9, 4.12 and 4.15 (Figure 4.10).

In this case, all three performance metrics used before will be considered to analyze the best possible solution among the three layers. All the equations below help to understand the energy consumption, cost factor and delay value analysis in case scenario 2, with different interfaces and data rates.

$$E_{\text{total}(a)} = (E_{ce} + E_{Co} + E_{ce}) * V_{\text{data}} \qquad (4.7)$$

$$E_{\text{total}(b)} = (E_{\text{wifi}} + E_{co} + E_{\text{wifi}}) * V_{\text{data}} \qquad (4.8)$$

$$E_{\text{total}(c)} = (E_{\text{wifi-di}} + E_{\text{shuttle}} + E_{\text{wifi-di}}) * V_{\text{data}} \qquad (4.9)$$

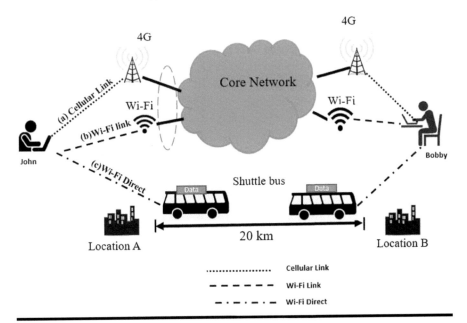

Figure 4.7 Crowdsensing data dissemination using existing public transport.

In Equations 4.7 through 4.9, energy consumption is calculated for all three options, where

E_{Co} = Energy consumed in the core network
E_{ce} = Energy intensity that amounts the total electricity consumption for the Internet and Internet traffic in the cellular link
$E_{wifi-di}$ = Energy consumed in the Wi-Fi link
$E_{wifi-di}$ = Energy consumed in offloading data onto the shuttle with Wi-Fi Direct as an interface
$E_{shuttle}$ = Energy consumed by the shuttle while traveling
V_{data} = Volume of data

$$D_{total(a)} = \frac{V_{data}}{B_{ce}} + \frac{V_{data}}{B_{co}} + \frac{V_{data}}{B_{ce}} \tag{4.10}$$

$$D_{total(b)} = \frac{V_{data}}{B_{wifi}} + \frac{V_{data}}{B_{co}} + \frac{V_{data}}{B_{wifi}} \tag{4.11}$$

$$D_{total(c)} = \frac{V_{data}}{B_{wifi-di}} + \frac{V_{data}}{B_{co}} + \frac{V_{data}}{B_{wifi-di}} \tag{4.12}$$

In Equations 4.10 through 4.12, the delay is calculated for the three options, where

B_{co} = Bandwidth used for core network
B_{ce} = Bandwidth value responsible for delay in cellular network
B_{wifi} = Bandwidth value used in Wi-Fi link using Wi-Fi interface
$B_{wifi-di}$ = Bandwidth value used in Wi-Fi Direct link
V_{data} = Volume of data

$$C_{total(a)} = (C_{ce} + C_{Co} + C_{ce}) * V_{data} \tag{4.13}$$

$$C_{total(b)} = (C_{wifi} + C_{co} + C_{wifi}) * V_{data} \tag{4.14}$$

$$C_{total(c)} = (C_{wifi-di} + C_{shuttle} + C_{wifi-di}) * V_{data} + C_{investement} \tag{4.15}$$

In Equations 4.13 through 4.15, the cost of the three options is calculated, where

C_{Ce} = Cost for transferring data through the cellular link to the core Internet network
C_{Co} = Cost of the core network
$C_{wifi-di}$ = Cost of the Wi-Fi link
$C_{wifi-di}$ = Cost of offloading the data onto the shuttle with Wi-Fi Direct as an interface

$C_{shuttle}$ = Minimal cost consumed by shuttle while traveling
V_{data} = Volume of data
$C_{investment}$ = One-time investment cost to install storage room on bus

Figure 4.8 shows that energy consumption is much lower when using the scheduled vehicle in comparison with the wired and wireless network options. In both case studies, Figures 4.9 and 4.10 illustrate the changes in costs and the delay values involved in the communication in each of these three layers, which shows that the cost involved in the case of the scheduled vehicle is lower than that for the wired and wireless network options.

The delay values have some threshold points to prove that the values will be acceptable as data volume increases, and they are less than those of other networks. Finally, all the performance metrics that satisfy that scheduled vehicle or public transport can be a better solution while combining with both of the traditional ways if data is delay tolerant.

4.3.3 Simulation Model and Analysis

Simulation and emulation are procedures often utilized for the study of wireless and wired systems. Contrasted to a test bed implementation, these strategies offer points of interest concerning scalability, reproducibility and cost-effectiveness. In the proposed research, simulation studies have been conducted using the ONE (Opportunistic Network Environment) Simulator [62]. ONE is an operator-based,

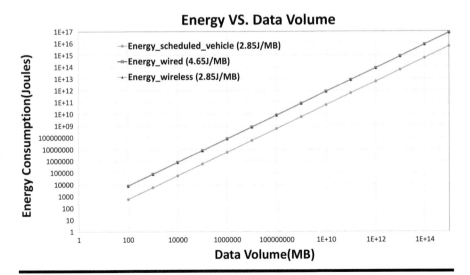

Figure 4.8 **Energy consumption in wired, wireless and scheduled vehicle networks.**

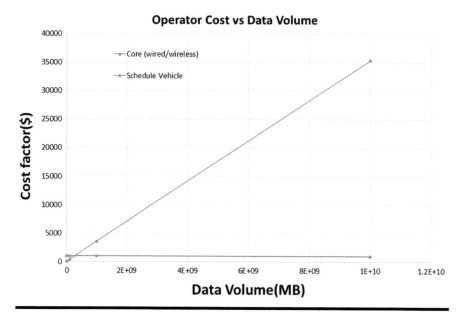

Figure 4.9 Cost factor included in wired, wireless and scheduled vehicle networks.

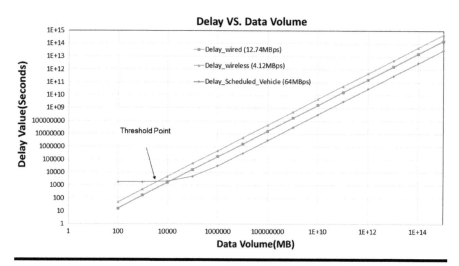

Figure 4.10 Delay value in wired, wireless and scheduled vehicle networks.

discrete-event simulation engine. The main functions of the ONE Simulator are the modeling of node movement, internode contacts, routing and message handling. Result collection and analysis are done with visualization, reports and post-processing tools. Once simulator scenarios are built, by defining simulated nodes and their capabilities, these parameters are entered in the default settings file.

We used the ONE Simulator to evaluate our results and check that our third layer or scheduled vehicles provide 100% delivery of messages, even in the case of delays. Therefore, we manage to sustain quality with big savings on energy consumption. We send crowdsensing data from two stationary nodes to another node at a fixed location. There are 12 buses scheduled on particular routes, and their transmission range is 1000 m with high-speed interface.

Experimental settings: The performance analysis of vehicles to carry data is done through the ONE Simulator. We used the Stationary Movement model for the source and destinations to communicate. The RoadSideUnit Placement model is used for RSU placement on the outside ring of the network to work as a wired network communication system. Buses move on a particular path using a Bus Movement model. We simulated three different scenarios to check the performance of a vehicle as a part of the communication network. According to changes in TTL values, we analyzed delivery probabilities and delay values in comparison with the RSU network, scheduled vehicle and random moving nodes. We first set the delivery capability of the RSU as 1, and the source originates the message and forwards it to RSU, rather than buses carrying the data. Next, we set the delivery capability at 0.5 for RSU and 1 for bus. In this option, data are forwarded to the bus and buses carry the data from the source to the destination.

Simulation parameters: Table 4.3 shows the all-simulation parameter to be used for simulation in ONE Simulator.

Evaluation results: We evaluate our result considering two performance metrics: delivery probability and delay value.

Table 4.3 Simulation Parameters

Simulator	ONE Simulator
Routing protocol	Unified router
Simulation time	5000s
Number of nodes	3
Numbers of vehicles	12
Interface	Wi-Fi, Bluetooth
Mobility model	Bus Movement, RoadSideUnit Placement, Stationary Movement
Node buffer	25K
Message TTL	5, 10, 30, 60, 90
Message size	500–1K
Delivery capability	0.5, 1

1. **Delivery probability:** Delivery probability expresses the fraction of the total number of generated messages that are successfully delivered. As Figure 4.11 shows, as the TTL values continue to increase, the delivery probabilities decrease and messages are dropped. But still, wireless nodes and scheduled vehicles have better delivery in comparison with the RSU network. As wireless networks and scheduled vehicles both work on the principle of DTNs, they both get the opportunity to forward messages to the destination.
2. **Delay value:** The average delay values are measured as the average time between the creation of each message and its delivery time. In Figure 4.12, as

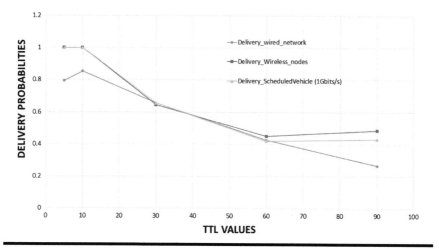

Figure 4.11   Delivery probabilities versus TTL values.

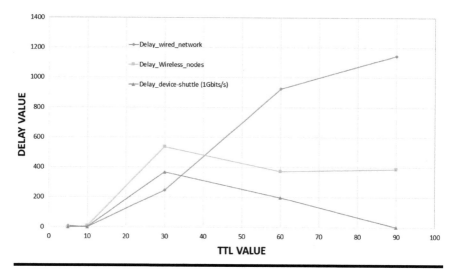

Figure 4.12   Delay values versus TTL value.

the TTL values increase, the delay decreases after an increment in the case of a scheduled vehicle. In a wired network, it is much more than wired networks and vehicles. In this case, scheduled vehicles can be considered a better solution even in terms of delay values.

4.4 Conclusions and Future Work

In this chapter, we analyzed the various perspectives, standpoints and modes of communication of traditional networks. Many routing and forwarding techniques have been used to alleviate network congestion and make communications more efficient. A literature briefing focuses on energy and network congestion reductions with good delivery probabilities, but with limited work using crowdsensing data dissemination techniques in a heterogeneous network scenario while keeping a good QoS and energy efficiency. We present a crowdsensing data dissemination technique in all three layers (wired network, wireless network and scheduled vehicle network) of a heterogeneous network, which aims to utilize the moving ability of public transport as a local crowd to reduce energy consumption. Many researchers have already discussed the concept of vehicular networks used as data carriers, but this chapter combines all the networks, such as moving crowd and the local crowd with cellular and public vehicles, to use and switch according to the requirements in terms of energy savings. This strategy can be used in the case of rural areas where wired and wireless networks are not available. To promote this strategy as the best solution for crowdsensing data dissemination with delays, incentive mechanisms can be introduced as part of our future work. The moving crowd can be incentivized with either monetary or nonmonetary rewards to use scheduled vehicles as a crowdsensing data dissemination medium as they are already placed (and running) and could be utilized for energy efficiency.

References

1. Cisco Visual Forecast Cisco visual networking index: Global mobile data traffic forecast update 2009–2014. Cisco Public Information, February 2010, p. 9.
2. DeHondt, J. Lean productivity enhancements and waste elimination through emerging technology. Gene Fliedner Decision and Information Sciences Department, School of Business Administration, Oakland University, 2015.
3. Bolcskel, H., et al. Fixed broadband wireless access: State of the art, challenges, and future directions. *Communications Magazine, IEEE*, 2001. **39**(1): pp. 100–108.
4. Asada, G., et al. Wireless integrated network sensors: Low power systems on a chip. In *Solid-State Circuits Conference, 1998, ESSCIRC'98, Proceedings of the 24th European*, The Hague, The Netherlands, 1998.
5. Heinzelman, W.R., J. Kulik, and H. Balakrishnan. Adaptive protocols for information dissemination in wireless sensor networks. In *Proceedings of the 5th Annual ACM/IEEE International Conference on Mobile Computing and Networking*, Seattle, WA, 1999.

6. Boukerche, A. and A. Darehshoorzadeh. Opportunistic routing in wireless networks: Models, algorithms, and classifications. *ACM Computing Surveys (CSUR)*, 2015. **47**(2): p. 22.

7. Penttinen, A. *Green Networking—A Literature Survey.* Helsinki: Aalto University, Department of Communications and Networking, 2012.

8. Chen, T., et al. Network energy saving technologies for green wireless access networks. *IEEE Wireless Communications*, 2011. **18**(5): pp. 30–38.

9. Han, C., et al. Green radio: Radio techniques to enable energy-efficient wireless networks. *IEEE Communications Magazine*, 2011. **49**(6): pp. 46–54.

10. Zhang, D., et al. 4W1H in mobile crowd sensing. *IEEE Communications Magazine*, 2014. **52**(8): pp. 42–48.

11. Yang, Z., C. Wu, and Y. Liu. Locating in fingerprint space: Wireless indoor localization with little human intervention. In *Proceedings of the 18th Annual International Conference on Mobile Computing and Networking*, Istanbul, Turkey, 2012.

12. Wang, Y., J. Zheng, and N. Mitton. Delivery delay analysis for roadside unit deployment in vehicular ad hoc networks with intermittent connectivity. *IEEE Transactions on Vehicular Technology*, 2016. **65**(10): pp. 8591–8602.

13. Hasan, Z., H. Boostanimehr, and V.K. Bhargava. Green cellular networks: A survey, some research issues and challenges. *Communications Surveys & Tutorials, IEEE*, 2011. **13**(4): pp. 524–540.

14. Hansen, J., et al. Climate change and trace gases. *Philosophical Transactions of the Royal Society of London A: Mathematical, Physical and Engineering Sciences*, 2007. **365**(1856): pp. 1925–1954.

15. Wang, X., et al. A survey of green mobile networks: Opportunities and challenges. *Mobile Networks and Applications*, 2012. **17**(1): pp. 4–20.

16. Kelly, T. and S. Head. ICTs and climate change. Technical Report. ITU-T Technology, 2007.

17. Cisco visual networking index: Global mobile data traffic forecast update. Cisco, 2010 [cited 2010–2015].

18. Sabhanatarajan, K. and A. Gordon-Ross. A resource efficient content inspection system for next generation smart NICs. In *Computer Design, 2008 (ICCD 2008), IEEE International Conference on*, Lake Tahoe, CA, 2008.

19. Costa, G.D., et al. The green-net framework: Energy efficiency in large scale distributed systems. In *Parallel & Distributed Processing, 2009 (IPDPS 2009), IEEE International Symposium on*, Rome, Italy, 2009.

20. Blackburn, J. and K. Christensen. A simulation study of a new green bittorrent. In *Communications Workshops, 2009 (ICC Workshops 2009), IEEE International Conference on*, Dresden, Germany, 2009.

21. Singh, S., et al. Energy efficiency in wireless networks—A composite review. *IETE Technical Review*, 2015. **32**(2): pp. 84–93.

22. Lilien, L., et al. Opportunistic networks: The concept and research challenges in privacy and security. In *Proceedings of the WSPWN*, Miami, FL, 2006, pp. 134–147.

23. Fall, K. A delay-tolerant network architecture for challenged Internets. In *Proceedings of the 2003 Conference on Applications, Technologies, Architectures, and Protocols for Computer Communications*, Karlsruhe, Germany, 2003.

24. Boldrini, C., M. Conti, and A. Passarella. Context and resource awareness in opportunistic network data dissemination. In *World of Wireless, Mobile and Multimedia Networks (WoWMoM 2008), 2008 International Symposium on a*, Newport Beach, CA, 2008.

25. Chen, L.-J., et al. A content-centric framework for effective data dissemination in opportunistic networks. *Selected Areas in Communications, IEEE Journal on*, 2008. 26(5): pp. 761–772.
26. Dhurandher, S.K., et al. Performance evaluation of various routing protocols in opportunistic networks. In *GLOBECOM Workshops (GC Wkshps)*, Houston, TX, 2011.
27. Jain, S., K. Fall, and R. Patra. Routing in a delay tolerant network. *ACM SIGCOMM Computer Communication Review*, 2004. 34(4), pp. 145–158.
28. Papadimitratos, P. and Z.J. Haas. Secure message transmission in mobile ad hoc networks. *Ad Hoc Networks*, 2003. 1(1): pp. 193–209.
29. Vahdat, A. and D. Becker. *Epidemic routing for partially connected ad hoc networks*. Technical Report CS-200006. Duke University, 2000.
30. Pelusi, L., A. Passarella, and M. Conti. Opportunistic networking: Data forwarding in disconnected mobile ad hoc networks. *Communications Magazine, IEEE*, 2006. 44(11): pp. 134–141.
31. Spyropoulos, T., K. Psounis, and C.S. Raghavendra. Spray and wait: An efficient routing scheme for intermittently connected mobile networks. In *Proceedings of the 2005 ACM SIGCOMM Workshop on Delay-Tolerant Networking*, Philadelphia, PA, 2005.
32. Spyropoulos, T., K. Psounis, and C.S. Raghavendra. Single-copy routing in intermittently connected mobile networks. In *Sensor and Ad Hoc Communications and Networks, 2004 (IEEE SECON 2004), 2004 First Annual IEEE Communications Society Conference on*, Santa Clara, CA, 2004.
33. Lindgren, A., A. Doria, and O. Schelén. Probabilistic routing in intermittently connected networks. *ACM SIGMOBILE Mobile Computing and Communications Review*, 2003. 7(3): pp. 19–20.
34. Soares, V.N., F. Farahmand, and J.J. Rodrigues. A layered architecture for vehicular delay-tolerant networks. In *Computers and Communications, 2009 (ISCC 2009), IEEE Symposium on*, Sousse, Tunisia, 2009.
35. Gorcitz, R.A., et al. *Vehicular Carriers for Big Data Transfers* (poster). In *IEEE Vehicular Networking Conference (VNC)*, 2012, pp. 109–114.
36. Cheng, N., et al. Vehicular WiFi offloading: Challenges and solutions. *Vehicular Communications*, 2014. 1(1): pp. 13–21.
37. Ott, J. and D. Kutscher. Drive-thru Internet: IEEE 802.11b for "automobile" users. In *INFOCOM 2004, Twenty-Third Annual Joint Conference of the IEEE Computer and Communications Societies*, Hong Kong, China, 2004.
38. Eriksson, J., H. Balakrishnan, and S. Madden. Cabernet: Vehicular content delivery using WiFi. In *Proceedings of the 14th ACM International Conference on Mobile Computing and Networking*, San Francisco, CA, 2008.
39. Balasubramanian, A., R. Mahajan, and A. Venkataramani. Augmenting mobile 3G using WiFi. In *Proceedings of the 8th International Conference on Mobile Systems, Applications, and Services*, San Francisco, CA, 2010.
40. Whitbeck, J., et al. Relieving the wireless infrastructure: When opportunistic networks meet guaranteed delays. In *World of Wireless, Mobile and Multimedia Networks (WoWMoM), 2011 IEEE International Symposium on a*, Lucca, Italy, 2011.
41. Mayer, C.P. and O.P. Waldhorst. Offloading infrastructure using delay tolerant networks and assurance of delivery. In *Wireless Days (WD), 2011 IFIP*, Niagara Falls, Canada, 2011.
42. Rebecchi, F., et al. Data offloading techniques in cellular networks: A survey. *IEEE Communications Surveys & Tutorials*, 2015. 17(2): pp. 580–603.

43. Dimatteo, S., et al. Cellular traffic offloading through WiFi networks. In *2011 IEEE Eighth International Conference on Mobile Ad-Hoc and Sensor Systems*, Valencia, Spain, 2011.

44. Bejerano, Y. Efficient integration of multihop wireless and wired networks with QoS constraints. *IEEE/ACM Transactions on Networking (TON)*, 2004. **12**(6): pp. 1064–1078.

45. Lei, H. and C.E. Perkins. Ad hoc networking with mobile IP. *ITG FACHBERICHT*, 1997: pp. 197–202.

46. Jönsson, U., et al. MIPMANET: Mobile IP for mobile ad hoc networks. In *Proceedings of the 1st ACM International Symposium on Mobile Ad Hoc Networking & Computing*, Boston, MA, 2000.

47. Tseng, Y.C., C.C. Shen, and W.T. Chen. *Mobile IP for mobile ad hoc networks:* An Integration and Implementation Experience. *IEEE Computer*, 2003. 36(5), pp. 48–55.

48. Laoutaris, N. and P. Rodriguez. Good things come to those who (can) wait. In *Proceedings of ACM HotNets*, Calgary, Canada, 2008.

49. Whitbeck, J. and V. Conan. HYMAD: Hybrid DTN-MANET routing for dense and highly dynamic wireless networks. *Computer Communication*, 2010. **33**(13): pp. 1483–1492.

50. Matis, M., et al. An enhanced hybrid social based routing algorithm for MANET-DTN. *Mobile Information Systems*, 2016. **2016**: p. 12.

51. Johnson, D., Y. Hu, and D. Maltz. The dynamic source routing protocol (DSR) for mobile ad hoc networks for IPv4 (No. RFC 4728), 2007.

52. Nguyen, H.A., S. Giordano, and A. Puiatti. Probabilistic routing protocol for intermittently connected mobile ad hoc network (PROPICMAN). In *2007 IEEE International Symposium on a World of Wireless, Mobile and Multimedia Networks*, Espoo, Finland, 2007.

53. Atzori, L., et al. The social Internet of things (SIOT)—When social networks meet the Internet of things: Concept, architecture and network characterization. *Computer Networks*, 2012. **56**(16): pp. 3594–3608.

54. Mayer, C.P. and O.P. Waldhorst. Where the Network ends: Infrastructure and delay tolerant networks in a hybrid future internet. In *6. GI/ITG KuVS Fachgesprach Future Internet*, Hannover, Germany, November 2010.

55. Le, V.-D., H. Scholten, and P. Havinga. Unified routing for data dissemination in smart city networks. In *Internet of Things (IOT), 2012 3rd International Conference on the*, Wuxi, China, 2012.

56. Le, V.-D., J. Scholten, and P. Havinga. Evaluation of opportunistic routing algorithms on opportunistic mobile sensor networks with infrastructure assistance. *International Journal on Advances in Networks and Services*, 2012. **5**(3–4): pp. 279–290.

57. Komnios, I. and V. Tsaoussidis. CARPOOL: Connectivity plan routing protocol. In *International Conference on Wired/Wireless Internet Communications*, Paris, France, 2014.

58. Komnios, I. and E. Kalogeiton. A DTN-based architecture for public transport networks. *Annals of Telecommunications*, 2015. **70**(11–12): pp. 523–542.

59. Komnios, I., F. Tsapeli, and S. Gorinsky. Cost-effective multi-mode offloading with peer-assisted communications. *Ad Hoc Networks*, 2015. **25**: pp. 370–382.

60. Schien, D., et al. The energy intensity of the Internet: Edge and core networks. In *ICT Innovations for Sustainability*, Lorenz Hilty and Bernard Aebischer (Eds.), pp. 157–170. Springer: Cham, Switzerland, 2015.

61. Iosifidis, G., et al. Enabling crowd-sourced mobile Internet access. In *IEEE INFOCOM 2014—IEEE Conference on Computer Communications*, Toronto, Canada, 2014.

62. Keränen, A., J. Ott, and T. Kärkkäinen. The ONE simulator for DTN protocol evaluation. In *Proceedings of the 2nd International Conference on Simulation Tools and Techniques*, Rome, Italy, 2009.

Chapter 5

Reputation Management on D2D Ecosystems

Dimitris Chatzopoulos, Pan Hui,
and Gunnar Karlsson

Contents

5.1 Device-to-Device Ecosystems

The popularity of smartphones is continuously increasing, with almost 1.5 billion of them sold every year,* while their capabilities exceed those of conventional servers of the previous 5–10 years. This proliferation of resourceful mobile devices with multiple sensors and network interfaces gave birth to another ecosystem where mobile devices assist each other on the execution of demanding tasks (task offloading), on reaching remote resources (traffic offloading), on retrieving contextual information (context awareness), and others. This is called the device-to-device (D2D) ecosystem and is composed of mobile devices that are able to communicate without the support of any fixed infrastructure. In D2D ecosystems, mobile users form mobile ad hoc networks (MANETs); communicate wirelessly via Wi-Fi Direct, Bluetooth, or even near-field communication (NFC); move unpredictably; and form temporary and delay-tolerant networks (DTNs). D2D ecosystems are challenging due to (1) their unpredictability, which is caused by users' mobility; (2) the limited, compared with conventional computer, computational resources and battery; and (3) the incentives required to motivate mobile users to participate. Figure 5.1 depicts a D2D ecosystem where mobile devices serve each other on miscellaneous tasks.

5.1.1 Applications on D2D Ecosystems

Applications on D2D ecosystems can be of many types. Traditional packet forwarding and routing in DTNs will regain popularity on the arrival of 5G technologies because it allows users' traffic to be routed via other proximal mobile devices. Moreover, new smartphones will be equipped with more than one cellular transceiver and will be able to connect with multiple networks at the same time. In another direction, applications will be able to be executed in more than one mobile device, following the paradigm of computation offloading, which was initially proposed for mobile cloud computing architectures. A multitude of offloadable applications have been proposed in literature during recent years [1]. A few D2D examples are peer-to-peer (P2P)-based k-anonymity location privacy [2], cooperative streaming [3], face recognition [4], video compression [5], and sensing [6]. The main difference between applications that have been introduced for MANETs and the aforementioned ones is the variety in the possible requested help. Packet-forwarding-like applications evaluate the help of each node only by whether it forwards or routes the packets it receives toward the destination. D2D applications have a computation offloading part that should be the main component in the used metrics. In other words, in traditional applications, all the mobile users have the same role and usually the same needs, while in modern D2D applications mobile users can diversify in many ways. Devices with different capabilities and

---

* https://www.statista.com/statistics/263441/global-smartphone-shipments-forecast/

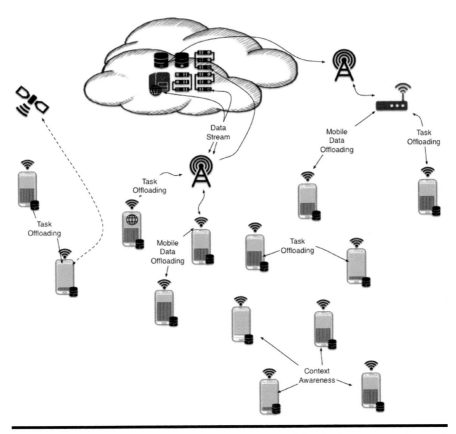

Figure 5.1 **In the considered D2D ecosystem, mobile devices can assist each other on various tasks and use remote services in parallel.**

variety of mobile applications are two of them. The two most popular mobile operating systems, Android and iOS, use object-oriented programming languages for development: Java and Objective C, respectively. In the context of mobile application offloading, an application can be seen as the union of two sets of classes: the offloadable and the nonoffloadable classes. In the runtime, the application offloading process will split the application into tasks and select the proper neighboring devices for offloading.

When a device receives a task from another node, it needs to allocate additional resources in order to process the task. Context-aware applications require help from other nearby devices in order to estimate the context. For example, mobile crowd sensing applications make use of devices' sensors to perform local measurements and share their data with each other. Context is a multifunctional variable of time and the ambient conditions and is a type of information that is worth sharing between mobile devices, regardless of whether they have past interactions. Computationally demanding applications like video compression require CPU cycles and memory, while traffic offloading applications ask for the helping devices to act as relays and help them access Internet resources. The use of these resources will cost the device in terms of battery and, in the case of data plan sharing, money. The requirements of the offloadable tasks can vary based on the functionalities of the application. These costs can be expressed as a function of the needed resources and the network overhead. Moreover, it is expected for mobile users to be cautious in installing and using a D2D offloading framework, considering that executing tasks for others implies high energy consumption. Not only that, but considering the selfish human nature [7], every user would be interested in extending his battery life by (1) trying to offload as many tasks as possible to others and (2) not accepting offloading requests from other users.

5.1.2 Hidden Market Design

Resource sharing has to be transparent from the user and needs to respect some sharing constraints (i.e., if the battery level is more than L% and the CPU utilization is less than U%, a Dalvik virtual machine instance, with XYZ characteristics can be initialized). Hidden market design principles allow mobile users, who are concerned about their devices' recourses, to impose a set of sharing bounds that determine the conditions for accepting an offloaded task via a graphical user interface. Such an interface has to follow the principles of the hidden market design, which states that (1) the complexities of the system must be hidden to the final user and (2) the user interface experience must be seamless. Based on these principles, we have implemented a user interface for Android, which is presented in Figure 5.2. Via this interface, mobile users can activate or deactivate the service, select the resources they are willing to share, and enforce the lower bounds for each resource. For example, if the battery lower bound is set to 30%, the device stops accepting tasks for execution from others when this battery level is reached.

5.1.3 Interactions between Mobile Users and Data Collection

Given that mobile users are self-interested, their goal is to utilize as many resources as possible from their neighbors while not sharing theirs. For that, incentive schemes and cooperation-enforcing mechanisms are needed to motivate users to share their available resources. Depending on the design of the cooperation mechanism, extra processing overhead and accounting messages are needed in order to maintain and share information related to mobile users' serviceableness in the whole ecosystem. The implementation of D2D architectures dictates the functionality of each component and its requirements. The aforementioned D2D applications and frameworks do not have any cooperation-enforcing and bookkeeping mechanism, and their neighbor selection mechanism (NSM) relies only on information related to the current situation of the devices. However, under the realistic case of self-interested mobile users, both cooperation-enforcing mechanisms and bookkeeping

Figure 5.2 A user interface proposed in [8] that is based on the principles of hidden market design.

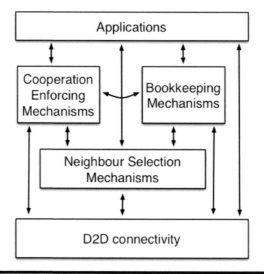

Figure 5.3   An application framework for D2D ecosystems.

mechanisms are needed in order to provide proper information to the NSMs. There are two main types of cooperation-enforcing mechanisms: credit-based ones and reputation-based ones. Both of them are associated with bookkeeping mechanisms that can be either centralized, which means that they require a centralized authority (server), or decentralized. Decentralized bookkeeping mechanisms are of two types; the first type requires a tamper-proof module that cannot be hacked, and the other has no such requirement but requires many messages to be exchanged between the mobile users. Figure 5.3 shows the basic components of such D2D offloading architecture and the interactions between them.

We define as interaction between two mobile users (2) the service and (2) the message exchange between them. Each interaction is associated with a number of messages that are exchanged before and after the actual service exchange. The messages that are exchanged before the service exchange are related to the characteristics and needs of the service (e.g., a D2D video compression application may require a Dalvik virtual machine with *XYZ* characteristics, while a context-aware application may need the readings of the gyroscope and the accelerometer for 1 minute). Furthermore, the mobile devices may also exchange data related to their own state (battery level, CPU utilization, connectivity conditions to the cellular network, etc.) in order to decide whether a service exchange would be beneficial. By the end of the service exchange, the mobile devices can exchange data to characterize the service exchange and can share these data with other nearby devices too. It is worth mentioning that the devices do not need to agree on the characterization of the service exchange. In order for the mobile devices to make efficient decisions regarding the service exchange with their neighbors, they need to collect data from past actions and evaluate their trustworthiness. Such data are related to the performance of other devices and their mobility. Given that it is inefficient to store data from every interaction, a mechanism for processing and evaluation is needed. Reputation systems process past interactions, evaluate devices' serviceableness, and provide reputation scores for each device.

5.1.4 Motivating Example

In Figure 5.4, we consider a mobile user named Bob who has around him four mobile users named Alice, Carol, David, and Eve. Bob just started an application that can be assisted by other nearby devices. Bob's device cannot ask for help from Eve because she is not in his coverage area. Alice is also not a candidate because she does not have the resource-sharing mechanism activated at that time. Carol and David are both accessible from Bob and have their resource-sharing mechanism ON. However, what Bob does not know about David is that he will not share his resources with him because the current utilization of his resources does not satisfy his sharing bounds. Bob's device will consider Carol's and David's devices. The collected data for Carol and David allow Bob's device to calculate their reputation and the expected connectivity and estimate how probable is to get the offloaded application parts back. Depending on these estimations and the quality of experience (QoE) or service guarantees that were imposed by the application developer, Bob's device will make the offloading decision.

5.2 Reputation Systems in D2D Ecosystems

The emerging-sharing economy increases the importance of trust in P2P marketplaces and services. Reputation is defined as the opinion that someone has about someone or something, or how much respect or admiration someone or something receives, based on past behavior or character. Reputation systems allow users to rate each other in communities in order to build trust through reputation. Reputation-based schemes discourage misbehavior by estimating users' reputation and punishing the ones with bad behavior. These schemes are based on the past attitude of the users, and each user calculates her trust to every other user. Trust and reputation are two concepts that get used so often that they get confused with each other. In terms of a mobile user, this means that someone is trustworthy if his or her actions are almost always what you expect an ideal user to do. Someone who is not trustworthy will frequently deviate from your expectations. In short, trust is your ability to accurately predict another person's behavior. On the other hand, reputation is not a prediction of the future, but knowledge of the past. Reputation is a memory tied to a specific identity. It is a collectively agreed upon version of how history has taken place. A strong reputation builds trust. Next, we present the most popular reputation and trust-based schemes, and in the next section we discuss how they can be adapted in D2D ecosystems.

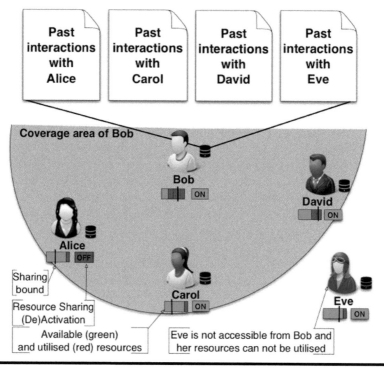

Figure 5.4   Example of a D2D ecosystem.

5.2.1 *Proposed Reputation Systems*

According to [9], trust is the subjective probability by which one individual expects another individual to perform a given action. Trust is useful only in an environment characterized by uncertainty and where the participants need to depend on each other to achieve their goals. Trust is context sensitive and subjective, while the formation of an opinion about someone's trustworthiness depends not only on the behaviors of the subject but also on how these behaviors are perceived by the agent. Moreover, trust is unidirectional and may not be transitive. Michiardi and Molva proposed CORE, a reputation system where a central authority keeps a record of everyone's cooperative behavior [10]. Users' reputation score can be between −1 and 1, and if it drops below 0 for one user, then it is classified as a *misbehaving entity*, and if she requests help the central authority will deny her request. The drawback of [10] is the assumption that a central authority has direct access to the mobile nodes and is aware of the whole system. CONFIDANT [11] also offers a punishment mechanism that isolates misbehaving users by not serving their requests. When a neighbor's reputation falls below a predefined threshold, service provision to the misbehaving user is interrupted. In such a way, there is no advantage for a user to misbehave because any resource utilization will be forbidden. The CONFIDANT protocol generates some additional traffic for reputation propagation. Malicious users may perform an attack by sending false alarms about other users and producing extra traffic. Another idea about reputation estimation is proposed by the authors of [12], where they decouple a peer's reputation as a service provider from its reputation as a service recommender, making the reputation more robust to malicious peers. Moreover, [13] presents a Bayesian network–based trust model and a method for building reputation based on recommendations in P2P networks. Since trust is multifaceted, peers need to develop differentiated trust in different aspects of other peers' capability.

The authors of SecuredTrust [14] analyze the different factors related to evaluating the trust in a multiagent system and propose a comprehensive quantitative model for measuring such trust. According to them, the overall trust depends on satisfaction, similarity, feedback credibility, direct trust, indirect trust, recent trust, historical trust, expected trust, decay model, and deviation reliability. Similarly, SORT [15] presents distributed algorithms used by peers to reason about the trustworthiness of other peers based on past interactions and mutual recommendations. Moreover, SORT evaluates interactions and recommendations based on importance, recentness, and peer satisfaction parameters. The recommender's trustworthiness and confidence about a recommendation are considered while evaluating recommendations.

The work presented in [16] introduces a trust model for the application of message forwarding and routing in MANETs. Every user has an initial budget of trust, and whenever a user detects a misbehaving user, she reports it and her trust budget changes accordingly. In such a way, a secure path between mobile users can be established. In the same direction, the authors of [17] present a trust domain–based security architecture for MANETs that copes with false disseminated information. Moreover, [18] proposes PowerTrust, a system that collects locally generated peer feedbacks and aggregates them to yield the global reputation scores and use a trust overlay network to model trust relationships among peers.

The authors of [19] proposed a solution aimed at detecting and avoiding misbehaving users through a mechanism based on a watchdog and a reputation system. The watchdog identifies misbehaving users by performing neighborhood monitoring: it observes the behavior of neighbors by promiscuously listening to the communications of users in the same transmission range. According to collected information, the reputation system maintains a value for each observed node that represents a reputation of its behavior. However, the watchdog approach has some important issues:

1. It is not clear how the proper devices will be selected to perform this role.
2. If the selected devices are predetermined, then they are a central point of failure.
3. It is not straightforward how the watchdogs are able to properly monitor the neighborhood.

4. Most importantly, it is not trivial how a watchdog can evaluate the help one user gave to another user or the harm it caused. This task may be easy in a packet-forwarding case, but in a generalized D2D ecosystem it is very difficult.

The authors of [20] deal with trust from the social point of view and propose protocols for packet forwarding in a social mobile setting. They use the fact that friend mobile users meet with high frequency in order to tolerate selfish behavior. In the next section, we discuss whether these presented works can be functional in D2D ecosystems.

Existing approaches in mobile computing, different from the ones presented for MANETs, consider trust a concept associated with authorities that pose no threat to an ecosystem and perform their tasks in a predefined way. In most of the cases, trusted authorities are responsible for signing data and providing cryptographic tasks. However, in a D2D ecosystem with no stable or secure connection to remote trusted servers, the concept of trust has to be relaxed. Considering reputation a set of evidence regarding the contribution of a user to her peers, trust is a random variable that depends on this evidence. Every mobile device calculates a trust score of every other device depending on her reputation. Moreover, given that the devices are heterogeneous and, depending on the users' configuration, cannot be helpful to every possible task, the trust scores should be application dependent and based on the collected evidence (i.e., the reputation) of other devices for every kind of asked help.

5.2.2 Bookkeeping of Reputation in D2D Ecosystems

At the end of each interaction between mobile devices, both parties are able to evaluate the interaction and, based on the implemented cooperation-enforcing mechanism, act accordingly. Each mobile device has to either report the outcome of the interaction to a remote server or store it locally. Moreover, it may only need to update the existing evaluation and then discard the receipt or store the evaluation in order to use it as part of historic data in the calculation of the user's score. Next, we present the possible ways to manage data from cooperation-enforcing mechanisms. Any centralized approach of a cooperation-enforcing mechanism requires a remote server for the bookkeeping [21]. There exist both synchronous and asynchronous approaches. In the first case, the mobile devices need to be connected to the server during the time of the association with their neighbors and submit their receipts after the end of the association. In the asynchronous case, the mobile devices can submit their receipts any time after the association. In a D2D ecosystem with mobile users, such a synchronous approach is functional if the mobile users have access to the remote server through a cellular connection. However, the cellular connection may impose significant latency whenever a mobile user wants to access stored data to estimate the score of another user. If the cooperation-enforcing mechanism is decentralized and there is no central server, the produced data need to be stored on the mobile devices. Such an approach lies in the area of MANET databases [22]. Since data availability in such networks is affected by the mobility and power constraints of the mobile users, the data need to be replicated. A number of data replication techniques have been proposed for MANET databases [23,24]. Users' mobility causes dynamic partitions, and while data replication may not exist for an entire network, it may be possible to maintain it in disjoint partitions within the network. The end result is that the database stored at each user may not be consistent with one another. As database updates are made, not all users are guaranteed to receive the updates within a reasonable time. The dynamic nature of MANETs makes maintaining the consistency of the data a challenge because multiple versions of the same information may exist throughout the network. When portions of the network become separated for a time, keeping data accurate may become impossible. We can identify power consumption, real-time requirements of applications, and network partitioning due to mobility as well as frequent disconnection as the most important issues to be considered in the design of a data replication technique for D2D ecosystems. These occurrences prolong transaction execution time due to the unavailability of remote resources. Replication techniques that predict the occurrence of network partitioning and replicate data items accordingly ahead of time are called partition-aware techniques. An ideal replication strategy for D2D ecosystems would be power aware, real-time aware, and partition aware. It is important to mention that the latency in the data access in MANET databases can be worse than the one in the remote server because it is probable for a user to not have all the required data locally in order to calculate the score of another user. If the cooperation-enforcing mechanism is credit based, the MANET database can impose functional problems in the mechanism because it will require many messages in order to guarantee consistency, if possible. If the cooperation-enforcing mechanism is designed to cover large-scale areas, the existence of a centralized authority is essential. More importantly, any MANET database should be installed to the mobile users with their consideration, and it is not easily justifiable why a mobile user would be interested in storing others' data in order to support a cooperation-enforcing mechanism for D2D ecosystems. This fact makes the MANET database one of the available services in the D2D ecosystem, and users that store its data should get rewarded whenever they respond to queries. This is can be an implementation of a cryptocurrency* in MANETs.

Depending on the design of the cooperation mechanism, extra processing overhead and accounting messages are needed in order to maintain and share information related to mobile users' serviceableness in the whole ecosystem. A lightweight reputation system can provide enough information to NSMs on D2D ecosystems in terms of

---

* Cryptocurrency: A digital currency in which encryption techniques are used to regulate the generation of units of currency and verify the transfer of funds, operating independently of a central bank.

1. Message exchanging
2. Processing requirements
3. Storage needs

In order to justify our argument, we define the price of inconsistency as the overhead caused to the mobile users by not selecting the most suitable helpers due to lack of complete information. Additionally, we discuss the cost of synchronization, which depends on the number of messages that need to be spread in order to inform every mobile user in the ecosystem and the required storage to save all the received evaluations that lead to complete information. We consider recommendation as a service, which is taking place whenever a mobile user is sharing her experience(s) with other mobile users. Based on the past interactions with the user that is giving the recommendations and her trustworthiness, her recommendations are evaluated. We formulate trust as a random variable, which depicts how much a user trusts another user on helping her with her a task. The calculation of trust is erroneous because a user cannot be familiar with all the interactions of all the other users. The price of inconsistency depends on this error. It is worth a reminder that any credit-based cooperation-enforcing mechanism has to guarantee that the price of inconsistency is zero. On the other hand, in such a mechanism, many more messages should have been exchanged. NSMs find the most suitable nearby users based on the score of each candidate. These scores are produced using collected data. These data are created either by the NSMs themselves or by other cooperation evaluation mechanisms. Usually, an application that is suitable for D2D ecosystems has a module to evaluate the contribution of each helper. Depending on the application and the request type, the evaluation can be based on the benefits caused by either the help or the saved costs. Our argument is based on the fact that each mobile device does not need to share all the data produced by its evaluation with others. On the other hand, depending on the context and the current knowledge of each mobile user, experience sharing between mobile users can be helpful to them to form opinions about others. There exists a three-way trade-off, as shown in Figure 5.5, between the required size of data that lead to a robust estimation about nearby devices (K), the amount of data that should be broadcasted to everyone whenever two mobile users interact (N), and the freshness of the stored data (mf).

Let us assume that the offloadable part of an application A_u is split into smaller tasks and the device of mobile user u has decided to ask the device of user v to help her with the task A_u^v. We formulate every possible application A as a combination of services. The set of all possible services is denoted by $S = S_1, S_2, \ldots, S_{|S|}$, and then each application is a vector in the power set of S, $A \in 2^S$.

Every mobile user, via a simple interface like the one we presented in Section 5.1.3, shares some of her resources. The set of shareable resources is denoted by $R = R_1, R_2, \ldots, R_{|R|}$, and there is a direct mapping from an application vector to the set of the minimum required resources in order for this application to be executed properly. Without loss of generality, we use normalized, to 1, values for the resources and the services.

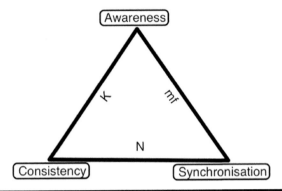

Figure 5.5 Three-way trade-off between consistency, awareness, and synchronization in D2D reputation systems.

Example 1: In the video compression application, one device connects with another one. It sends a video, and then the helping device compresses the video, and later when they meet again, the helping device sends the compressed video to the initial device. This means that the application requires S_1 = computation and S_2 = local network connection. These two services are mapped to R_1 = CPU, R_2 = Memory, and R_3 = TCP socket with the nearby device.

Example 2: In the cooperative streaming application, multiple devices are connected with each other and some of them are connected to the Internet and download parts of the same video and share these parts with each other. The cooperative streaming application requires $S\{2\}$ = $ local network connection, S_3 = Internet connection, and S_4 = network buffer. These three services are mapped to R_1 = CPU, R_2 = Memory, R_3 = TCP socket, and R_4 = mobile database.

So $A = (a_1, a_2, \ldots, a_{|S|})$ can be described by the set of services it is based on, and we assume that there exists a mapping function $r()$ such that

$$r(A) \rightarrow R$$

At the end of each interaction, both devices are able to evaluate the interaction, and based on the implemented cooperation-enforcing mechanism they will act accordingly. We define a history matrix on each user u for user v, $H_u^v \ \varepsilon \ I^{\wedge K, |S|+1}$, where I is the unit interval and K is the number of interactions that v has saved. The number of columns of H_u^v are two more than the number of services because each row contains in the second to last column the ID of the mobile device that shared the stored entry with user u about user v. If this device is u herself, then the value is equal to zero. The last column keeps the timestamp of the interaction, while the ith column stores the evaluation of the ith service that user v promised to provide.

The recommendation is one of the $|S|$ services, which takes place whenever a mobile user shares her experience(s) with other mobile users. Based on the past interactions with the user that is giving the recommendations and her trustworthiness, her recommendations are evaluated. The way H_u^v will be used, as well as the value of K, depends on the cooperation-enforcing mechanism. We consider the case of using a combination of a trust and a reputation system. We refer to trust using the following definition: "Trust is the ability to accurately predict another person's behavior." We formulate trust as a random variable $\theta_u^v(A_u^v)$, which depicts how much user u trusts user v on helping her with her application part A_u^v. Given that A_u^v can be mapped to a set of minimum required resources and that u is not familiar with all interactions of v with the remaining mobile users, we argue that $\theta_u^v(A_u^v)$ is erroneous. Moreover, in the case where u had access to all the stored passed data with v's interactions with other users, she could have built a more robust estimation of $\theta_u^v(A_u^v)$. We define $\varphi_u^v(A_u^v)$ as the trust score u could have built about v if she had access to all v's interactions (K = infinity). All these interactions can be known to u if the cooperation-enforcing mechanism is credit based; then the enforced integrity guarantees would have allowed u to be familiar with v's interactions. On the other hand, in such a mechanism, many more messages would have been exchanged. Then the price of inconsistency is given by the absolute difference between $\varphi_u^v(A_u^v)$ and $\theta_u^v(A_u^v)$.

5.3 An Asynchronous Reputation System for D2D Ecosystems

In [25] we proposed a lightweight reputation system for D2D ecosystems that works in a distributed way in each mobile user independently and without any need for coordination. Our approach is based on the use of the first and second moments of $\theta_u^v(A_u^v)$. For the calculation of $\theta_u^v(A_u^v)$, we select the framework of beta distribution. In order to find $\theta_u^v(A_u^v)$, we need to first calculate the parameters of beta distribution, which are $\alpha_u^v(A_u^v)$ and $\beta_u^v(A_u^v)$. $\alpha_u^v(A_u^v)$ is the weighted sum of all the positive interactions u has collected about v for all cases where the services of A_u^v were used, while $\beta_u^v(A_u^v)$ is the weighted sum of the negative ones. The weights in these sums are the trust score of the mobile user that provided the entry. If the entry was provided by u, the weight equals 1. In order to calculate these two parameters, we only need H_u^v and no communication with other devices.

We consider recommendations as one type of service in the D2D ecosystems. Any new coming mobile user does not have any collected data for the other mobile users. Whenever a mobile user u has in her neighbor list a candidate for help v with empty H_u^v, she assumes that $\alpha_u^v(A_u^v) = \beta_u^v(A_u^v) = 1$, which gives v a trust score of 0.5 with a uniform distribution and the highest possible variance $\alpha_u^v(A_u^v)$. We assume that every mobile user has a confidence score (i.e., maximum acceptable $\sigma_u^v(A_u^v)$) in her opinion about other mobile users, and in order to satisfy this confidence score, she requests information about others from other trusted friends. Given that the information that is produced by our proposal is going to be used by NSMs that are aiming to improve the QoE of D2D applications, it is important to not marginalize mobile users for their selfish attitude in the past. On the other hand, free riders should not have the same confrontation as the altruists. During the calculation of $\alpha_u^v(A_u^v)$ and $\beta_u^v(A_u^v)$, we use a multiplication factor on each entry i of H_u^v:

$$mf = \lambda \big/ \left(t - t_i \right)$$

where λ is a positive tuning parameter, t is the current timestamp, and t_i is the timestamp for when entry i was collected. mf slows down the decrease of the variance and feeds the need for new entries. In the general case, any mobile user requests others' recommendation whenever her current evaluation has bigger variance than the imposed threshold. If her current entries is less than K, she just enters more entries in her history matrix. If her history matrix is full, she discards entries with small contribution to the distribution. The contribution of each entry is calculated by multiplying the entry by mf times the mean of the trust distribution of the mobile user who offered the entry. If the entry was inserted by the user herself, we multiply only by mf. Next, we present a static analysis of users that are uniformly distributed and show how their population size and the connectivity between them affect the number of messages needed to maintain the integrity of a credit-based system. Also, we show how the number of available data from past interactions affects the consistency of the users regarding the serviceability of the others.

We produce instances of static random geometric graphs using MATLAB™. We distribute users uniformly in a [0, 1] × [0, 1] area. In Figure 5.6 we show how many retransmissions are required in order for one message to arrive to all the users of the network in the case of 1000 or 2000 users. The x-axis of the two leftmost figures

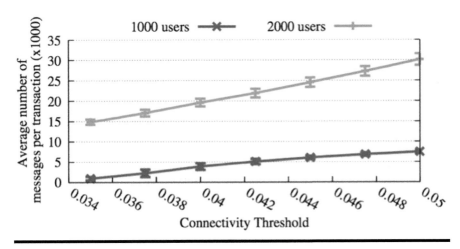

Figure 5.6 Depending on the used D2D technology for the interconnectivity of the users, the number of messages needed to synchronize the users varies.

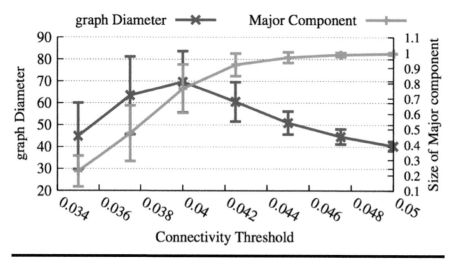

Figure 5.7 Depending on the used D2D technology for the interconnectivity of the users, the diameter and size of the major connected component of the connectivity graph vary.

shows the connectivity threshold. By connectivity threshold, we define the ratio of the coverage radius of a smartphone to the whole examined area. For the simulation purposes, we assumed that all the smartphones have the same coverage radius and are uniformly distributed in a squared area. Moreover, Figure 5.7 shows how the graph diameter is decreasing and the fraction of the users in the major connected component is converging to 100$% when the connectivity threshold is increasing.

Figure 5.8 shows how the amount of the stored interactions, K, affects the diversity of the trust scores between mobile users. We have randomly selected a

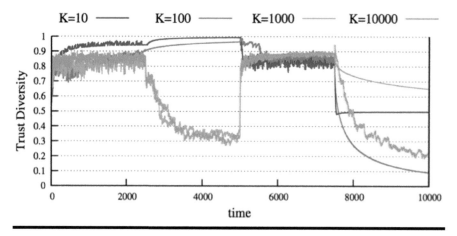

Figure 5.8 Depending on the available storage for the collected data from recommendations, the adaptability of our algorithm to user profile changes varies.

probability for each user to be helpful to others in the begging of the simulations, which last for 10,000 slots, and on every quarter of the simulation time we change the profile of the users to either not helpful or completely altruistic (helping everyone). We used $\lambda = 1$ for the calculation of *mf*. As we can see from the plot, if the collected data are not enough, the trust estimation cannot follow the profile change.

5.4 A Reputation Middleware for D2D Ecosystems

In [26], we proposed OPENRP, novel, lightweight, and scalable system middleware that provides a unified interface to D2D applications. When an application wants to perform a D2D task, it delegates the task to the middleware, which takes care of choosing the best peers to collaborate with and sending the task to these peers.

5.4.1 OPENRP Architecture

OPENRP evaluates and updates the reputation of participating peers based on their mutual opportunistic interactions. As depicted in Figure 5.10, OPENRP is modular and is composed of independent daemon components that communicate with each other using interprocess messages. The main components of the system are responsible for accepting requests from applications, collecting information about the surrounding neighbor devices, collaborating with close by peers, and evaluating the reputation of the collaborators. As we can see from Figure 5.10, applications are totally unaware of the underlying details of the system and interact with the middleware through a properly specialized API layer, which is presented later. Moreover, OPENRP provides a configuration interface that the final user can use to select the resources she is willing to share when participating in the crowd computing collaboration system.

5.4.2 API of OPENRP

OPENRP provides an extensive API to frameworks and D2D applications to be used for D2D collaboration. The main purpose of the middleware is to receive tasks from the above applications and delegate them to nearby devices. Figure 5.8 depicts the basic interactions between mobile devices when using OPENRP. We show the sequence diagram of a classic scenario of collaboration between two devices, where Application i running on device dev1 wants to send a task to another device. To keep the diagram clean, we omit the arguments of the API's methods, leaving only their names. Application i on dev1 uses the request() command to delegate a task to OPENRP. When OPENRP finds an available nearby device, dev2 in our example, it sends a request for collaboration, encoded as REQ, and waits for the response RESP from the other device. If dev2 decides to collaborate, represented by the YES branch in the diagram of Figure 5.9, Application i in dev1 is notified by OPENRP through the ack() method. At this point, Application i on dev1 registers with OPENRP to be notified when the eventual response of the task is ready. Then, OPENRP on dev1 sends the task to the OPENRP on dev2 using the message TASK. The middleware on the second device, dev2, receives the task and uses the method process() to delegate it to the appropriate application, which is the same Application i as the one running on dev1 that knows how to handle the task.

After the application processes the task, it uses the method answer() to inform OPENRP on dev2, which then sends a message ANS to the requesting device dev1 that the task was correctly processed. OPENRP on dev1 then uses the method response() to pass the eventual result to Application i or to simply inform it that the task was processed by dev2, depending on the task type. Application i uses the method feedback() to advertise its experience of the collaboration with device dev2. OPENRP collects this information and uses it to update the reputation of dev2. If for some reason device dev2 refuses to process the task (the reason can be

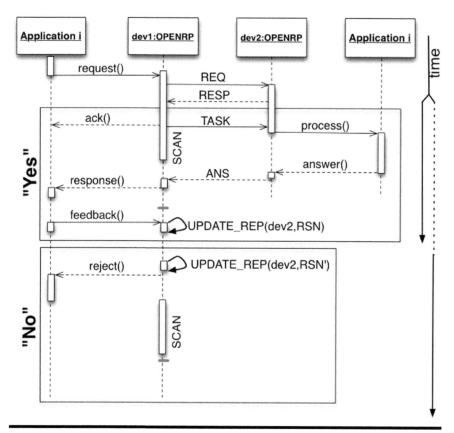

Figure 5.9 **The use of the API from Application i on device dev1 asking to collaborate with Application i on device dev2.**

included in the response message RESP), OPENRP in dev1 updates the reputation score about dev2, informs Application i using the method reject() that the other device did not accept the task, and keeps scanning for other devices. Apart from the methods described in the previous example, OPENRP provides the applications with more API commands, as listed here:

- getNmostTrusted(N,S|A) returns the N most trusted neighbors. This method has both a synchronous and an asynchronous version, returning the currently nearby devices or all the known ones, respectively.
- sync()forces OPENRP to contact the neighbors and ask for updates.
- getUserRep(id) returns the reputation of a specific user.
- getListOfSensors(dev d) gives the list of the available sensors for use at device d.
- getMeetingFrequency(dev d) returns the meeting frequency between the current device and device d.
- getAvgContactDuration(dev d) returns the average contact duration between the current device and device d.
- abortTask() is used by an application to notify the middleware that the task is not needed anymore.
- cacheResult(task, timestamp) is used by an application to ask the middleware to cache a result so that for future calls of the same task there will be no need to process the task.
- clearCache() is used by an application to ask the middleware to remove all its cached task results from the cache.
- removeFromCache(task) is used by an application to ask the middleware to remove a specific task from the cache.
- estimateExecution() is used by an application to estimate the needed time or energy to execute one task given the current conditions of the device.
- getCurrentConditions(dev d) is used by an application to get the current battery level, CPU utilization, and other available resources of device d.

5.5 A Synchronized Reputation System for D2D Ecosystems

Synchronized systems guarantee data coherence and integrity. In a synchronized reputation system for D2D ecosystems, all the mobile devices should assign the same trust scores to the rest of the participating devices. Apart from MANET databases, as discussed in Section 5.2.2 on blockchains, the underlying technology of cryptocurrencies can offer equivalent functionality with higher security guarantees.

The popularity of digital currencies, especially cryptocurrencies, has been continuously growing since the appearance of Bitcoin [27]. Bitcoin is a P2P cryptocurrency protocol enabling transactions between individuals without the need for a trusted authority. Its network is formed from resources contributed by individuals known as miners. Users of Bitcoin currency create transactions that are stored in a specialized data structure called a blockchain. Bitcoin's security lies in a proof-of-work scheme, which requires high computational resources at the miners. These miners have to be synchronized with any update in the network, which produces high-data-traffic rates. A blockchain is a decentralized and distributed digital ledger that is used to record transactions across many communicating parties so that the record cannot be altered retroactively without the alteration of all subsequent blocks and the collusion of the network. Each transaction contains information that is related to the sender and the receivers, and apart from credits it can also contain extra information.

Despite advances in mobile technology, no cryptocurrencies have been proposed for mobile devices. This is largely due to the lower processing capabilities of mobile devices compared with conventional computers and the poorer Internet connectivity compared with that of the wired networking. In [28], we proposed LocalCoin, an alternative cryptocurrency that requires minimal computational resources, produces low data traffic, and works with off-the-shelf mobile devices. LocalCoin replaces the computational hardness that is at the root of Bitcoin's security with the social hardness of ensuring that all witnesses to a transaction are colluders. It is based on opportunistic networking rather than relying on infrastructure and incorporates characteristics of mobile networks, such as users' locations and their coverage radius, in order to employ an alternative proof-of-work scheme. LocalCoin features (1) a lightweight proof-of-work scheme and (2) a distributed blockchain.

Decentralized cryptocurrencies have to deal with three main challenges:

1. Proof of ownership—Users should be able to prove that they have the amount of money they claim to have.
2. Double-spending avoidance—A defense mechanism against double spending. (Users are not able to spend the same money more than once.)
3. Incentives—For the stakeholders.

5.5.1 Proof of Ownership

LocalCoin uses a lightweight storage architecture by extending the concept of blockchain in a distributed fashion, where each user can store as many blocks as she wants Figure 5.11. The proposed distributed blockchain has a redundancy factor between the users. LocalCoin, similarly to Bitcoin, stores transactions in blocks. All the transactions in the same block are collectively verified. BS denotes the size of each block. In order for one block to be created, a minimum number of users to verify each transaction, denoted by mVu, is needed (i.e., at least BS·mVu users are informed about each transaction on one block). The relationship of these variables with the total amount of users affects the time needed to verify one block and prove the ownership of all the users that own these transactions (Figure 5.11).

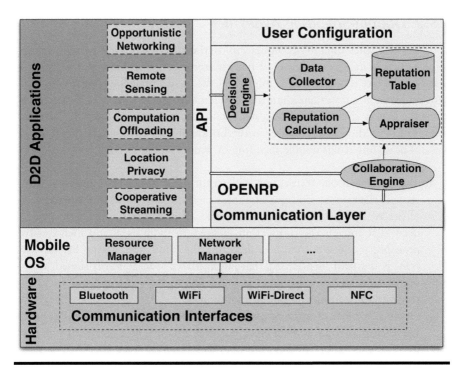

Figure 5.10 Architecture of OPENRP, as placed on top of a mobile OS and below D2D applications.

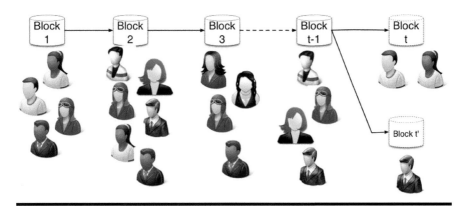

Figure 5.11 Distributed blockchain. Every block is stored to more than one but not every user. Users who own at least one transaction in a block have to store it.

5.5.2 Double-Spending Avoidance

LocalCoin nullifies Bitcoin's computation overhead via incorporating a novel protocol, which is designed for the ad hoc environment. Bitcoin's proof of work is based on the fact that cheating is improbable because a malicious user has to solve hard problems at a faster rate than the total remaining users. In LocalCoin, cheating is made very difficult because a malicious user has to misinform the majority of a set of trusted users. Every user in the LocalCoin protocol selects the users she trusts. LocalCoin avoids double spending in two ways:

1. The receiver of one transaction will accept the transaction if and only if she receives the transaction signed by at least a minimum number of trusted users of her trusted network, denoted by mTr. This constraint imposes a useful delay that spreads the transaction message to more users and increases the probability of one trusted user detecting the same input to another transaction. It is worth mentioning that any initiated transaction is signed by the sender, and we assume that it is impossible for a malicious user to fake a transaction by pretending to be another user.
2. During the block creation process, every participant checks for double-spending attempts. To avoid fake block creation attempts by a set of collaborative malicious users, LocalCoin enforces the average distance between the users that will verify the creation of a new block , denoted by aVd, to be more than a predefined value. This last constraint allows the block creation messages to be scattered to as many users as possible.

5.5.3 Incentives

LocalCoin extends the transaction fee schema in order to motivate mobile users to participate. Transaction fees motivate users to forward messages, and block fees motivate them to store as many blocks from the distributed blockchain as possible. Transaction fees are important because mobile users are competing for them and they broadcast any received transaction. Every transaction includes an amount of LocalCoins that are collected by the mobile user who will first inform the receiver of the transaction about the transaction. Block fees are important because users store the created blocks in order to be able to verify the creation of new ones. Whenever a block is created, the mobile users that verified each transaction because they were aware of it share the LocalCoins that were included in these transactions as block fees.

5.6 Conclusion and Future Work

In this chapter, we initially presented D2D ecosystems and listed some popular applications. Next, we indicated the importance of hidden market design in the sharing of the users' resources without the need for their intervention, and we discussed the types of the interactions between the mobile users. After that, we provided an asynchronous reputation system and discussed its characteristics in terms of storage, message exchange, and robustness, and we presented a three-way trade-off. We continued with the presentation of OPENRP, a reputation middleware that offers an API to application developers that want to implement D2D applications. Finally, we discussed LocalCoin, a cryptocurrency that can be used as a synchronous reputation system for D2D ecosystems.

We expect future work to be on the cut of reputation systems and cryptocurrencies. There is a continuously increasing interest in the field of cryptocurrencies from both industry and academia. Multipurpose blockchains, like the one developed by Ethereum, will be part of our everyday life and reputation systems will be deployable on top of cryptocurrencies.

Acknowledgments

This chapter is supported by the Sponsorship Scheme for Targeted Strategic Partnership (SSTSP) of the Hong Kong University of Science and Technology (HKUST).

References

1. F. A. Silva, G. Zaicaner, E. Quesado, M. Dornelas, B. Silva, and P. Maciel. Benchmark applications used in mobile cloud computing research: A systematic mapping study. *Journal of Supercomputer*, vol. 72, no. 4, pp. 1431–1452, April 2016.
2. R. Shokri, G. Theodorakopoulos, P. Papadimitratos, E. Kazemi, and J. P. Hubaux. Hiding in the mobile crowd: Location privacy through collaboration. *IEEE Transactions on Dependable and Secure Computing*, vol. 11, no. 3, pp. 266–279, May 2014.
3. L. Keller, A. Le, B. Cici, H. Seferoglu, C. Fragouli, and A. Markopoulou. Microcast: Cooperative video streaming on smartphonesI. In *Proceedings of the 10th International Conference on Mobile Systems, Applications, and Services (MobiSys '12)*. New York: ACM, 2012, pp. 57–70.
4. R. Kemp, N. Palmer, T. Kielmann, and H. Bal. Cuckoo: A computation offloading framework for smartphones. In *Mobile Computing, Applications, and Services: Second International ICST Conference, MobiCASE 2010, Santa Clara, CA, USA, October 25–28, 2010, Revised Selected Papers*. Berlin: Springer, 2012, pp. 59–79.
5. D. Chatzopoulos, K. Sucipto, S. Kosta, and P. Hui. Video compression in the neighborhood: An opportunistic approach. In *IEEE ICC 2016 Ad-Hoc and Sensor Networking Symposium (ICC '16 AHSN)*, Kuala Lumpur, Malaysia, May 2016.
6. Y. Huang, A. Tomasic, Y. An, C. Garrod, and A. Steinfeld. Energy efficient and accuracy aware (E2A2) location services via crowdsourcing. In *2013 IEEE 9th International Conference on Wireless and Mobile Computing, Networking and Communications (WiMob)*, Rome, Italy, October 2013, pp. 436–443.

7. C. Bermejo, R. Zheng, and P. Hui. An empirical study of human altruistic behaviors in opportunistic networks. In *Proceedings of the 7th International Workshop on Hot Topics in Planet-scale mObile computing and online Social neTworking (HotPOST '15)*. New York: ACM, 2015, pp. 43–48.

8. D. Chatzopoulos, M. Ahmadi, S. Kosta, and P. Hui. FlopCoin: A cryptocurrency for computation offloading. *IEEE Transactions on Mobile Computing*, vol. PP, no. 99, p. 1.

9. H. Yu. Building robust crowdsourcing systems with reputation-aware decision support techniques. ArXiv e-prints, February 2015.

10. P. Michiardi and R. Molva. Core: A collaborative reputation mechanism to enforce node cooperation in mobile ad hoc networks. In *Proceedings of the IFIP TC6/TC11 Sixth Joint Working CCMS: ACMS*. Deventer, The Netherlands: Kluwer, BV, 2002, pp. 107–121.

11. S. Buchegger and J.-Y. Le Boudec. Performance analysis of the confidant protocol. In *Proceedings of the 3rd ACM International Symposium on Mobile Ad Hoc Networking & Computing, MobiHoc '02*. New York: ACM, 2002, pp. 226–236.

12. G. Swamynathan, B. Y. Zhao, and K. C. Almeroth. Decoupling service and feedback trust in a peer-to-peer reputation system. In *Parallel and Distributed Processing and Applications—ISPA 2005 Workshops*. Berlin: Springer, 2005, pp. 82–90.

13. Y. Wang and J. Vassileva. Trust and reputation model in peer-to-peer networks. In *Third International Conference on Peer-to-Peer Computing*. New York: IEEE, 2003, pp. 150–157.

14. A. Das and M. M. Islam. Securedtrust: A dynamic trust computation model for secured communication in multiagent systems. *IEEE Transactions on Dependable and Secure Computing*, vol. 9, no. 2, pp. 261–274, 2012.

15. A. B. Can and B. Bhargava. Sort: A self-organizing trust model for peer-to-peer systems. *IEEE Transactions on Dependable and Secure Computing*, vol. 10, no. 1, pp. 14–27, 2013.

16. Z. Liu, A. W. Joy, and R. A. Thompson. A dynamic trust model for mobile ad hoc networks. In *Distributed Computing Systems, 2004 (FTDCS 2004), Proceedings of the 10th IEEE International Workshop on Future Trends of*, May 2004, pp. 80–85.

17. M. Virendra, M. Jadliwala, M. Chandrasekaran, and S. Upadhyaya. Quantifying trust in mobile ad-hoc networks. In *Integration of Knowledge Intensive Multi-Agent Systems, 2005, International Conference on*, Waltham, MA, April 2005, pp. 65–70.

18. R. Zhou and K. Hwang. Powertrust: A robust and scalable reputation system for trusted peer-to-peer computing. *IEEE Transactions on Parallel and Distributed Systems*, vol. 18, no. 4, pp. 460–473, 2007.

19. S. Marti, T. J. Giuli, K. Lai, and M. Baker. Mitigating routing misbehavior in mobile ad hoc networks. In *Proceedings of the 6th Annual International Conference on Mobile Computing and Networking (MobiCom '00)*. New York: ACM, 2000, pp. 255–265.

20. A. Mei and J. Stefa. Give2get: Forwarding in social mobile wireless networks of selfish individuals. *IEEE Transactions on Dependable and Secure Computing*, vol. 9, no. 4, pp. 569–582, July 2012.

21. C. Curino, E. Jones, R. A. Popa, N. Malviya, E. Wu, S. Madden, H. Balakrishnan, and N. Zeldovich. Relational cloud: A database service for the cloud. In *5th Biennial Conference on Innovative Data Systems Research*, Asilomar, CA, January 2011, pp. 235–240.

22. L. D. Fife and L. Gruenwald. Research issues for data communication in mobile ad-hoc network database systems. *ACM SIGMOD Record*, vol. 32, no. 2, pp. 42–47, 2003.
23. P. Pabmanabhan and L. Gruenwald. Dream: A data replication technique for real-time mobile ad-hoc network databases. In *Data Engineering, 2006 (ICDE '06), Proceedings of the 22nd International Conference on*. New York: IEEE, 2006, p. 134.
24. P. Padmanabhan, L. Gruenwald, A. Vallur, and M. Atiquzzaman. A survey of data replication techniques for mobile ad hoc network databases. *The VLDB Journal: The International Journal on Very Large Data Bases*, vol. 17, no. 5, pp. 1143–1164, 2008.
25. D. Chatzopoulos and P. Hui. Asynchronous reputation systems in device-to-device ecosystems. In *Proceedings of the 8th ACM International Workshop on Hot Topics in Planet-Scale Mobile Computing and Online Social Networking (HotPOST '16)*. New York: ACM, 2016, pp. 25–30. doi: http://dx.doi.org/10.1145/2944789.2944873.
26. D. Chatzopoulos, M. Ahmadi, S. Kosta, and P. Hui. OPENRP: A reputation middleware for opportunistic crowd computing. *IEEE Communications Magazine*, vol. 54, no. 7, pp. 115–121, July 2016.
27. S. Nakamoto. Bitcoin: A peer-to-peer electronic cash system. 2009. (Online) available from: https://bitcoin.org/bitcoin.pdf.
28. D. Chatzopoulos, S. Gujar, B. Faltings, and P. Hui. LocalCoin: An ad-hoc payment scheme for areas with high connectivity: Poster. In *Proceedings of the 17th ACM International Symposium on Mobile Ad Hoc Networking and Computing (MobiHoc '16)*. New York: ACM, 2016, pp. 365–366.

Chapter 6

Network Connectivity and Data Quality in Crowd-Assisted Networks

İzzet Fatih Şentürk and Metin Bilgin

Contents

6.1 Introduction

The ubiquity of various sensors on pervasive computing devices has enabled observing the physical world, in real time, with a plethora of sensors. Thanks to human involvement, mobility, which is inherent to human-accompanied devices, makes

Table 6.1 Mobile Devices Equipped with Sensors

Sensor	iPhone 8 [1]	Samsung S8 [2]	Garmin Vivoactive 3 [3]	Tesla Model X [4]
GPS	✓	✓	✓	✓
Accelerometer	✓	✓	✓	
Barometer	✓	✓	✓	
Compass		✓	✓	
Gyro	✓	✓		
Proximity	✓	✓		✓
Ambient light	✓	✓		✓
Fingerprint	✓	✓		
Thermometer			✓	
Heart rate		✓	✓	
Iris		✓		
Hall		✓		
Camera	✓	✓		✓
Microphone	✓	✓		✓
Radar				✓

mobile crowd sensing (MCS) possible. Smartphones, wearable devices, and vehicular systems are some of the human-accompanied mobile devices with a variety of onboard sensors, as listed in Table 6.1. The wearables market is already diversified with a dazzling array of products for various applications, including entertainment, fitness, and medical. More than 400 different wearable devices are already available [5] to change the way we work, exercise, and interact. According to Gartner, the market trend indicates a 90% increase by 2021 in the worldwide wearable device sales [6].

In MCS, generated data is consumer-centric in the sense that a certain degree of user participation is essential at different stages of the application. The MCS process can be defined as a series of steps: task allocation [7–10], sampling, and data collection [11–13].

- ***Task allocation:*** In the first step, sensing tasks are defined and assigned to participants. Depending on the application, the number of participants may be crucial to provide a certain level of service quality. For instance, while a single participant is sufficient to monitor the movement pattern (i.e., transportation mode) of an individual for a personal health application, the phenomena at a larger scale (e.g., traffic congestion monitoring) require collective sensing of many individuals. Considering the overhead to be incurred to perform the assigned tasks and the privacy concerns due to revealing sensitive information such as location, people may be reluctant to participate in the system. Therefore, incentive mechanisms should be applied to attract more participants.

- **Sampling**: During the sampling phase, environmental conditions are observed through employed sensors. Sensors to be employed are subject to the sensing context defined by the sensing task. Apart from the sensing context, other requirements, such as location, time, and the sampling rate, are defined by the sensing task. However, fulfillment of the task request is contingent on the participant's approval. A major concern in this phase is the accuracy of the indicated value at the output of the employed sensor(s). Due to the heterogeneity of devices, accuracy may vary between sensors of different manufacturers. Besides, malicious participants may send manipulated data deliberately without performing the actual task.

  Based on the degree of user involvement in the sampling phase, crowd sensing can be classified into two categories: participatory crowd sensing and opportunistic crowd sensing. In participatory sensing, user involvement (participation) is explicit (i.e., human work is required to satisfy the request). On the other hand, sensing tasks are automated and the data is collected without active user involvement in opportunistic sensing. MCS can also be classified based on the data generation modes. Besides sensors, MCS can also leverage user-contributed data from social networks. However, this chapter focuses on mobile sensing data.

- **Data collection**: In the last step, sensor readings are collected by the remote server (i.e., data center). Different communication models can be applied based on the wireless communication methods provided to the MCS application by the participant. Note that, despite its availability, participants may not opt to use cellular data (e.g., LTE [14]) considering the communication cost that needs to be covered by individuals and the overhead on the battery. If long-range wireless communication means are provisioned, data can be forwarded to the remote server immediately. However, communication can be limited to Wi-Fi or Bluetooth as well. In such a case, despite progress in sampling, data collection will be postponed until the mobile device is connected to a network. Such limitations introduce delay in data collection, which may not be acceptable for some MCS applications. Delay also occurs when the data collection frequency is set lower than the sampling frequency in order to minimize the communication overhead.

User involvement in MCS not only poses challenges but also offers unprecedented opportunities. Unlike traditional sensor networks, which require deployment of custom hardware, crowd sensing applications leverage devices in sheer numbers that are already deployed in the field. This not only avoids the deployment cost of specialized sensing infrastructure but also minimizes the time required to launch the application. On the other hand, the control of the data quality is rather limited. Two major concerns are the sparsity of participants at a certain location at a given time and the quality of the generated data in terms of accuracy, integrity, and latency. Considering the fact that rational users control mobile devices, selfish users may be reluctant to participate in crowd sensing applications in order to conserve energy, storage, and computing resources. Furthermore, malicious users may generate fabricated data on purpose. Therefore, new methods must be developed to ensure a certain degree of data quality while increasing the number of participants by applying incentive mechanisms in conjunction.

MCS employs heterogeneous devices with diverse sensing capabilities, various wireless communication standards, and different mobility models. Besides mobile devices, MCS may also comprise stationary sensor nodes. In such a case, mobile devices can be used to improve some of the network performance metrics, such as connectivity and coverage. Network connectivity is a fundamental issue that needs to be tackled in wireless networks and a major concern of this chapter. If sensor nodes form a partitioned network, mobile devices in MCS can be exploited to provide intermittent connectivity between sensor nodes and the remote server. This is similar to MSNs that are intermittently connected through mobile data collectors (MDCs) [15,16]. The resulting network model is referred to as a crowd-assisted network (CaNET) and the mobile devices in the network signify corresponding participants. A sample CaNET can be found in Figure 6.1.

This chapter discusses how public crowds can be exploited to intermittently connect nodes in a partitioned network and enable a CaNET. First, we assume a sensor network with disjoint nodes such that none of the nodes are reachable from the rest of the network. Then we introduce humans to the network. Humans offer inherent mobility, and the accompanying devices provide a means of wireless communication. Humans, involved in the network, are regarded as participants. The availability of wireless communication combined with inherent mobility enables employing participants as MDCs. Participants collect data from sensor nodes within their proximity and relay the collected data to the remote server. This scheme provides intermittent connection to the nodes so that the data sampled in the network can be forwarded to the remote server. The resulting intermittently connected network is regarded as a CaNET. Considering the fact that the sampled data can be manipulated or even fabricated by malicious people, we introduce malicious participants into the network. Then we investigate data quality and define two different metrics, namely, accuracy and integrity, to evaluate the quality of the collected data. In order to resolve data conflicts and determine the actual data that is most likely sampled from individual sensors, we present two novel approaches based on arithmetic average and frequency. Both approaches are evaluated in terms of integrity and accuracy.

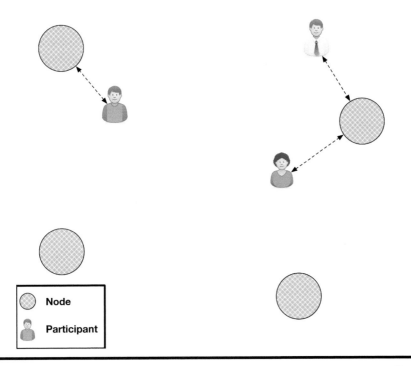

Figure 6.1 A sensor network with four disconnected nodes. Sensor readings cannot be delivered to the remote server due to the limited transmission range. Participants offer intermittent connectivity for otherwise disconnected nodes and enable a CaNET.

The rest of the chapter is organized as follows. We highlight the key differences between wireless (WSN) or mobile (MSN) sensor networks and MCS in Section 6.2. Connectivity and coverage issues in CaNETs are discussed in Section 6.3. Two different approaches to assess the reliability of the collected data are presented in Section 6.4. Proposed approaches are evaluated in Section 6.5. The chapter is concluded and open issues are discussed in Section 6.6.

6.2 From Sensor Networks to MCS

The availability of low-cost sensor nodes with wireless communication capabilities has enabled WSNs comprising sensors in large quantities to monitor an area of interest and track certain events or phenomena. MSNs have emerged with the deployment of mobile sensors to take advantage of mobility and sensing at the same time. Unless sensor nodes have inherent mobility capabilities, node mobility can be enabled by attaching nodes to mobile robots [17] as well. Considering the limited on-board batteries of the mobile nodes and excessive energy overhead of mobility compared with other network activities, such as messaging [18], mobility should be limited and employed in a controlled manner upon needed to extend the lifetime of the mobile. There are several solutions where mobility is employed as a means for optimizing the network performance in terms of connectivity [19], coverage [20], and lifetime [21,22]. Mobility is also exploited to tolerate node failures [23].

Compared with sensor networks where energy is constrained and usually non-rechargeable, MCS leverages human-accompanied devices, such as smartphones, wearables, and intelligent vehicles, which are less restricted in terms of power supply, computational power, communication capacity, memory, and storage. Furthermore, such devices are maintained by users through charging as needed, and they usually provide direct access to the Internet. Since the cost of ownership is addressed by its users, MCS is more cost-effective than MSNs. Also, the possibility of user involvement in large numbers offers a scalable and flexible solution that can be easily extended to span across large areas.

The financial cost of providing the same coverage with traditional sensor networks in large areas renders the application infeasible. For instance, a CO_2 monitoring application within the fifth ring in Beijing (about 900 km²) would require the deployment of at least 90,000 sensor nodes and around 1,000,000 relay nodes to maintain full area coverage and communication connectivity [24], which is undesirable. On the other hand, it is possible to provide 90% coverage for the same region by employing 6300 taxis [24].

In the MCS, mobility is inherently exploited through human mobility. This avoids the high mobility cost that exists in MSNs, which is much more than the messaging cost [18] and therefore should be controlled carefully. In MCS, on the other hand, human-accompanied devices with sensing, localization, and wireless communication capabilities collect samples whenever they are within the predetermined area to be monitored and report their readings to a remote server to enable large-scale sensing tasks. Human mobility not only improves coverage but also offers intermittent connectivity for otherwise disconnected nodes, as focused on in this chapter.

Several advantages of MCS exist over MSNs. The primary advantage is the involvement of humans to cover the cost of devices, handle mobility, take care of the communication cost, and maintain devices (e.g., recharging) to sustain their operations. Millions of smart devices and intelligent vehicles already exist, and they are ready to be employed around the world. The second advantage is the abundance of resources. Due to the limited form factor of the sensor nodes, WSNs are limited in terms of computation, communication, and energy. Most of the WSNs employ low-rate, short-range wireless technologies, such as IEEE 802.15.4 [25], to communicate within the network, and thus network-wide collaboration is critical to sustain connectivity with the base station (BS), which acts as a gateway between the network and the remote user. However, due to the depletion of limited on-board batteries and the exposure of the nodes to harsh environmental conditions, the network can be subject to random node failures. While some of the failures can be compensated with redundancy, failure of the cut-vertex nodes renders the network partitioned. Such problems do not exist, most of the time, for MCS applications since individuals have direct access to the Internet through one of the long-range wireless communication methods, such as LTE or Wi-Fi.

MCS also poses several challenges that need to be addressed. Unlike traditional sensor networks where the number of nodes and their locations are known in advance, controlling data quality is more challenging in MCS. Stability is a major concern typically. The number of participants is expected to fluctuate due to random user mobility. Eagerness to participate may also change based on the actual condition of the device, such as battery life and user preferences. The heterogeneity of devices is another challenge. In the ideal case, sensors are expected to produce the same output for the same input. However, sensor readings for the same environmental conditions can be different even for sensors from the same manufacturer depending on the sensor calibration. Ambient conditions surrounding the device and physical alignment with the phenomena to be monitored may also impact the sensor accuracy adversely. Consider an application where ambient noise is to be monitored. Sensor readings will be subject to the location of the device (e.g., hand, bag, pocket). Besides, some users may deliberately send false data to earn money without performing the assigned task. Spatial redundancy is another issue that both poses a challenge and offers opportunities. On the one hand, duplicate data from multiple participants must be eliminated. On the other hand, redundancy can be exploited to assess the reliability of the collected data. The idea is evaluating disparity in the data collected from the same region by different participants. For a detailed discussion, refer to Section 6.4. In any case, drastic measures must be taken accordingly to address the mentioned challenges and ensure data quality in terms of accuracy, latency, and integrity.

Another concern from the participants' perspective is privacy. Collected data may contain sensitive information. In general, sensor readings are tagged with location and time. Moreover, collected data can be analyzed to reveal patterns such as trajectories and extract sensitive information such as participants' home and office addresses [26]. Anonymization is an option to provide preservation. Providing anonymity, on the other hand, may encourage users to send incorrect data due to the complexity of taking action on anonymous users. Privacy in MCS is an open issue and new methods should be developed to ensure user privacy at a certain level.

Besides the risk of privacy issues, users also consume their own resources for data collection. The total cost of ownership includes the initial purchase price of the device, maintenance (e.g., charging the device as needed), mobility cost, and communication cost. To compensate the associated costs and improve participation, incentive mechanisms should be developed. Otherwise, users will be reluctant to participate. Several incentive strategies exist [27–30], which can be classified into entertainment, service, and money [27]. Besides such incentives, social recognition can be another motivating factor for participation. Some of the key features of MSNs and MCS are summarized in Table 6.2.

Table 6.2 Some of the Key Differences between MSNs and MCS

	Mobile Sensor Networks	Mobile Crowd Sensing
Operators	Institutions	Individuals
Network	Homogeneous network with a fixed network size	Dynamic network with diverse devices
Control	Autonomous control with a possible intervention	User controlled with limited or no access to the hardware
Maintenance	Self-organizing; energy is limited and usually not rechargeable	User maintained
Mobility	Limited autonomous movement	Inherent mobility with no control
Context	Limited to the deployed region and the employed hardware	Location and available sensors may change
Sampling and collection	Full control	User dependent
Scalability	Good	Best

6.3 Connectivity and Coverage in CaNETs: Challenges and Opportunities

Mobile crowds can be employed to provide intermittent connectivity to an otherwise disconnected network of stationary sensor nodes and enable a CaNET. This model offers unprecedented opportunities in network connectivity and coverage. In this section, we briefly describe the challenges regarding connectivity and coverage in MSNs first and then discuss how these issues are addressed by CaNETs.

In MSNs, limited energy supplies on the nodes enforce a limited transmission range to minimize the communication overhead. A limited transmission range, on the other hand, requires nodes to collaborate with each other in order to send their data to the BS. Therefore, connectivity of the sensor nodes with the BS must be maintained at all times in order to sustain network operations. However, nodes may fail arbitrarily for various reasons, such as battery depletion, hardware malfunction, or an external damage. Such failures may partition the network into disjoint subsets if the failed nodes are cut vertices. When a partition is isolated from the rest of the network, collected data within the partition cannot be delivered to the BS and the sensing coverage drops drastically.

Several solutions exist in the literature, which deals with the connectivity restoration problem through employing mobility. To restore the connectivity of a partition with the rest of the network, network topology should be adjusted accordingly. Three of the most common approaches are as follows:

- *Restructuring network topology through relocation of the existing mobile nodes*: Since mobility imposes significant energy cost to the limited batteries of the nodes, movement distance should be minimized. In addition, if the scope of the damage is too wide, determining the nodes to be relocated and their final locations is another challenge.
- *Deploying additional nodes between the partitions*: In the second approach, determining the minimum number of nodes to be introduced to ensure recovery is crucial. Multiple batches of deployment will be inevitable if the number of deployed nodes is not sufficient to guarantee connectivity. Besides, a self-configuring scheme is required to determine movement destinations of the nodes.
- *Employing mobile data collectors*: MDCs must be assigned to partitions uniformly in such a way that the tour lengths of MDCs are minimized and the load among MDCs is balanced.

It can be noticed that all solutions pose certain challenges, and the primary challenge in MSNs is limited energy. While the first two approaches provide a stable connection, the last one offers intermittent connectivity. Compared with MSNs, we assume sensor nodes to be stationary in CaNETs. Therefore, the first approach is not applicable to CaNETs. We also assume that additional nodes will not be deployed between the existing nodes to connect them. Thus, the second approach is not an option for CaNETs. On the other hand, the last solution is very similar to a CaNET. However, there are a few differences. First, mobility is inherent to participants with no additional cost in a CaNET. We also assume no control on the participants' mobility pattern. Therefore, unlike MSNs, we cannot optimize the movement path of a participant and balance the overhead between participants. The lack of control on the mobility simplifies the problem in CaNETs. On the other hand, success of the data transmission cannot be guaranteed due to the random mobility. Apparently, the number of participants and the size of the application area for mobility are crucial to the data delivery success. To demonstrate the correlation, we have evaluated network connectivity in CaNETs with a varying number of participants. The details and further discussion can be found in Section 6.5.

Despite limited energy supplies of MSNs, CaNETs are less restricted in terms of energy. First, most of the human-accompanied mobile devices have higher energy capacity than tiny sensor nodes. Furthermore, they can be easily recharged as needed. On the other hand, sensor networks often operate unattended in environments where human intervention is limited. Thus, recharging is not an option for MSNs typically. Also, the lack of control on the hardware and random human mobility render any mobility optimization algorithm inapplicable.

Collaboration in message passing, another challenge inherent to MSNs, is not a case most of the time in CaNETs. Such collaboration is not imposed in CaNETs since devices are equipped with long-range communication technologies such as LTE, which provides direct access to the rest of the world. Considering possible limitations on the employed communication method, we demonstrate a scenario where participants need to collaborate in order to forward their data, as illustrated in Figure 6.2a. Despite its availability, participants may not opt to leverage long-range communication means and impose limitations on available communication methods to be employed considering the communication overhead in terms of cost and energy. In the scenario given in Figure 6.2a, only low-rate, short-range wireless communication technologies are assumed to be available for the CaNET. Therefore, participants are employed as MDCs to relay data from nodes to the BS. The BS has direct Internet access and is employed as a gateway between the network and the remote server. This scheme requires participants to visit the BS periodically in order to forward the collected data. Assuming limited or no control on the mobility patterns of the participants, the data collection rate will decline drastically compared with the direct communication.

The reliability of the collected data is a major challenge in a CaNET. Malicious participants may alter samples they obtain and send falsified data on purpose. The situation can easily deteriorate further if participant collaboration is assumed, as in Figure 6.2a. Note that in collaboration networks, one malicious participant has the potential to manipulate other participants' data while relaying. In the case of malicious participants, the data center will likely receive conflicting data from different participants. If direct Internet access, is assumed as in Figure 6.2b, it can be possible to validate the data sampled from the corresponding sensor node. We present two different approaches in Section 6.4 for data validation. To demonstrate the correlation between the ratio of malicious participants and the data quality, we have evaluated the proposed data validation approaches in Section 6.5.

6.4 Approach

Due to sensor calibration or malicious participants, the data center may receive conflicting data from different participants. However, a certain level of data reliability is essential in order to meet application-level objectives. Therefore, we need a truth discovery model so that conflicting data can be resolved. In this chapter, we consider a model where data is sampled by sensor nodes and forwarded by participants. Sensor nodes are assumed to maintain a high level of accuracy such that individual nodes sustain producing the same output for the same input. This

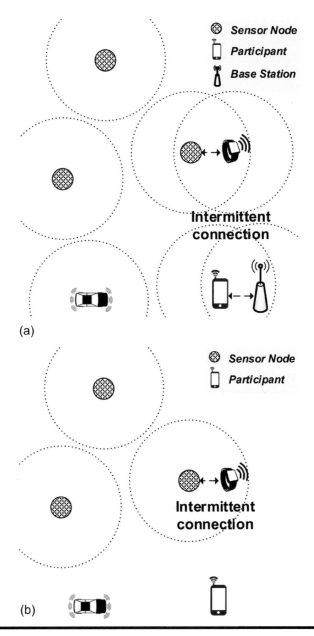

Figure 6.2 (a) Participants enable LTE to forward data obtained from sensor nodes. Participants do not have to collaborate and visit the BS since they have direct Internet access to forward the collected data. (b) Participants employ one of the short-range communication methods. Only the BS has direct Internet access. Participants should collaborate and visit the BS in order to forward the collected data.

provides repeatable measurements for the same environmental conditions. In the proposed model, we assume a sample space to represent the set of possible values that can be sampled from sensors. Each sensor node is assumed to report ambient conditions denoted with a value randomly selected from the sample space. Different sensors may report different values; however, we assume the sampled data to be invariant for individual sensors. As part of the threat model, we assume that malicious participants are manipulating the actual data obtained from sensors. Since participants are mobile, a single malicious participant may alter one or more sensors' data depending on the trajectory. The data center may receive true and false values for each sensor node. By analyzing the set of values corresponding to different nodes, we aim to determine the true value for each node by employing data validation methods.

In this section, we introduce two different data validation approaches, namely, arithmetic average–based data validation (AADV) and frequency-based data validation (FDV), to resolve conflicting data. By analyzing the collected data, we can determine the value that most likely sampled and improve data quality. We consider the connectivity issue as well in Section 6.5. But, apart from that, we introduce two different metrics, accuracy and integrity, to evaluate data quality. To clarify both metrics, let us consider two different scenarios where light sensors and gesture sensors are employed to obtain ambient light information and detect gestures (e.g., left to right, up to down), respectively. While precise measurements are required for the gesture sensor, some deviation from the actual value may still be acceptable for the light sensor. Thus, we classify sensors into two categories based on the reported data: incremental and particular. Considering this classification, we define accuracy and integrity metrics to evaluate sensors generating incremental and particular samples, respectively. The details can be found in Section 6.5.

Considering privacy concerns, we assume anonymization of the participants. Thus, the data center does not collect information regarding participants. The data center only collects sensor readings and the corresponding node ID where the data is sampled. Note that samples of a sensor node may not be received by the data center due to random mobility of the participants. On the other hand, some nodes may be visited multiple times by the same participant or various participants. The more the data is collected from a node, the greater is the chance of determining the actual data. The ratio of malicious participants is also crucial to the success of the data validation.

In the presented model, without loss of generality, five different values are assumed in the sample space: *A*, *B*, *C*, *D*, and *E*, as given in Table 6.3. Sample space can represent both incremental and particular data, such as *very low, low, medium, high*, and *very high* for ambient light levels or *left to right, right to left, up to down, down to up*, and *wave* for gesture directions. We assume participants to have direct Internet access to forward data, and therefore no collaboration is required between participants. AADV and FDV approaches are detailed next.

Table 6.3 Notations Used in the Algorithms

n_i	Sensor node $i \in N$
R_i	The set of data collected from n_i
v_j	Sample $j \in R_i$
P	The set of possible sensor data values: A, B, C, D, E
S	The set of actual data reported from corresponding nodes
A	The set of data determined by AADV for corresponding nodes
F	The set of data determined by FDV for corresponding nodes

6.4.1 Arithmetic Average–Based Data Validation

A set of conflicting data may be reported to the data center for the same node due to various reasons, such as malicious participants or the lack of sensor accuracy. To maintain data quality, we aim to resolve data conflict and determine the actual data sampled by the corresponding nodes. Thus, characteristics of the obtained data set must be described. One approach is to pursue quantitative methods and apply frequency analysis. The idea is to evaluate the number of occurrences of each data and reveal the central tendency of the overall data set. This approach enables us to represent the data set through a single value with the most accuracy. One of the common measures of central tendency is the mean value. While mean has various definitions depending on the context, we consider arithmetic average in this approach. AADV, as the name suggests, applies the arithmetic average of the data reported for corresponding nodes. Since numerical data is required for the mean, each data in the sample space is mapped to a numerical value starting, from 1 and incremented by 1. See Algorithm 6.1.

Algorithm 6.1: AADV (N, R)

```
1: for i = {1, 2, . . . , |N|} do
2:     R_i = getSensorReadings(n_i)
3:     if R_i == 0 then                    //No data from n_i
4:         continue
5:     end if
6:     sum = 0
7:     counter = 0
8:     for ∀v_j ∈ R_i do
9:         sum += v_j
10:        counter ++
11:    end for
12:    A_i = sum / counter
13: end for
```

6.4.2 Frequency-Based Data Validation

Mod is another popular measure of central tendency that can be applied to identify a single value to represent the whole data set with the most accuracy. The idea is to employ the probability density function to determine the data that is most likely sampled from corresponding nodes. Thus, we evaluate how frequently each data is reported. If the data appears more, its relative likelihood to be the actual data is assumed to be increased. FDV considers the mod of the data reported for corresponding nodes and sets the data that appears the most as the validated data. See Algorithm 6.2.

Algorithm 6.2: FDV (N, R)

```
1: for i = {1, 2, . . . , |N|} do
2:      R_i = getSensorReadings(n_i)
3:      if R_i = = 0 then                    //No data from n_i
4:          continue
5:      end if
6:      counts[] = 0
7:      for ∀v_i ∈ R_i do
8:          counts[ordinal(v_i)]++
9:      end for
10:     max = 0
11:     maxIndex = 0
12:     for j = {1, 2, . . . , |P|} do
13:         if counts[j] > max then
14:             max = counts[j]
15:             maxIndex = j
16:     end if
17:     end for
18:     F_i = maxIndex
19: end for
```

6.5 Experimental Evaluation

This section explains the experiment setup, performance metrics, and obtained results.

6.5.1 Experiment Setup

The efficiency and validity of the presented approaches are tested through simulations. We have considered stationary sensor nodes to monitor the surrounding physical phenomena and a remote server (i.e., data center) to collect and process the data. The nodes are deployed randomly in such a way that none of the nodes are reachable from the rest of the network. To provide intermittent connection between the nodes and the data center, mobile devices are introduced into the network. Mobile devices represent participants in the crowd, and the resulting intermittently connected network is regarded as a CaNET. A random-way-point mobility model is applied to mobile devices.

We have varied the number of sensor nodes (i.e., 4–10), the number of participants (i.e., 1–4), and the size of the application area (i.e., 200 m × 200 m – 500 m × 500 m) during experiments. For each setup, experiments were carried out 30 times and the average is reported for significance.

6.5.2 Performance Metrics

We have considered four different performance metrics for assessment.

- **Connectivity:** This metric reveals the number of nodes that were able to establish intermittent connection to the data center at least once. A higher number of connected nodes denotes improved network coverage.
- **Message count:** This metric indicates the total number of messages collected from the network and successfully delivered to the data center. Unlike the first metric, message count implies the duration of the network connection.
- **Accuracy:** This metric evaluates the deviation from the expected value. This metric is especially useful to assess data quality when sensors with incremental sample values (e.g., brightness level obtained from ambient light sensor) are employed.
- **Integrity:** This metric signifies whether the collected data is consistent with the generated data. This metric does not tolerate deviation and expects the same value with the actual data. This metric is essentially useful when precise measurements are needed (e.g., directions obtained from a gesture sensor).

Algorithm 6.3 elaborates how accuracy and integrity metrics are computed.

Algorithm 6.3: Evaluate_Quality (S, A, F)

```
1: a = 0, v = 0
2: for i = {1, 2, . . . , |S|} do
3:     if S_i = = A_i then
4:         v++
5:     else
6:         a += |S_i - A_i|
7:     end if
8: end for
9: Accuracy = a / |N|, Integrity = v / |N|
```

6.5.3 Performance Results

Figures 6.3 and 6.4 present the number of connected nodes with varying number of participants and network sizes, respectively. To observe the relation between the network density and the network connectivity, the size of the application area is varied between 200 m × 200 m and 500 m × 500 m. The number of sensor nodes is set to 10 in Figure 6.3. As Figure 6.3 suggests, network connectivity improves when the number of participants is increased. This is expected since participants follow a random movement pattern and introducing additional participants into the network increases the chance of an encounter with the sensor nodes.

However, the size of the application area impacts the network connectivity adversely. It can be observed from Figure 6.3 that the network connectivity suffers in networks with low node density, especially with a single participant. The number of connected nodes declines almost 55% when the size of the application area is increased from 200 m × 200 m to 500 m × 500 m in a network with one participant. On the other hand, for the same scenario, connectivity drops 11% when four participants exist in the network. It can be concluded that redundancy in mobility alleviates the adverse effects of the sparse networks.

In Figure 6.4, we employ a single participant and vary the number of nodes and the size of the application area. The results are relative to the node count. Figure 6.4 denotes

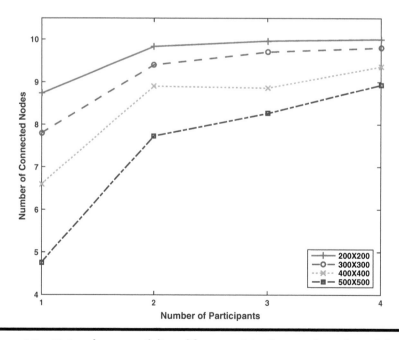

Figure 6.3 Network connectivity with respect to the number of participants and the size of the monitored application area. The number of nodes is set to 10, while the number of participants is varied between 1 and 4.

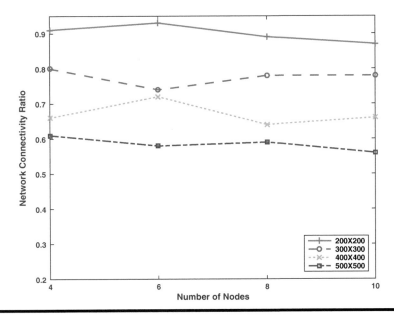

Figure 6.4   Network connectivity with respect to the number of nodes and the size of the monitored application area. The number of participants is set to 1, while the number of nodes is varied between 4 and 10.

that the number of connected nodes is proportional to the number of nodes in the network even when the application area varies in size. As expected, the network connectivity ratio improves when the network size declines. This is due to the decreased average distance between the nodes, which leads to an increased chance of visit by a participant.

The total numbers of messages successfully delivered to the data center are given in Figures 6.5 and 6.6 for varying numbers of participants and nodes, respectively. Figure 6.5 reveals that the number of delivered messages increases when the number of participants increases. The size of the application area is inversely proportional to the message count. This can be attributed to the increased average distance between the nodes in sparse networks and the decreased probability of participant–node encounter. As expected, the highest number of message count is attained when the application area is set to 200 m × 200 m.

Figure 6.6 suggests that the number of messages successfully delivered to the data center increases with the increased node count. This is expected due to the increased message count generated by the sensor nodes and the improved likelihood of a participant–node encounter considering the increased node density when the size of the application area is fixed. On the one hand, improvement in the message count diminishes in larger application areas. If the number of nodes is increased from 4 to 10 when the size of the application area is set to 200 m × 200 m, the message count increases 156%. On the other hand, if the size of the application area is set to 500 m × 500 m, then the increase in the message count declines to 51%.

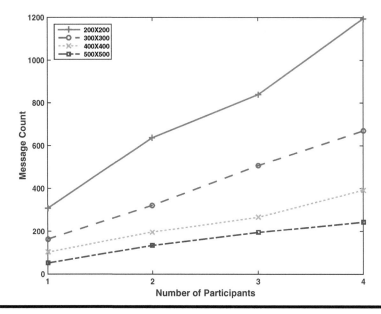

Figure 6.5 **The number of delivered messages with respect to the number of participants and the size of the monitored application area. The number of sensor nodes is set to 10, while the number of participants is varied between 1 and 4.**

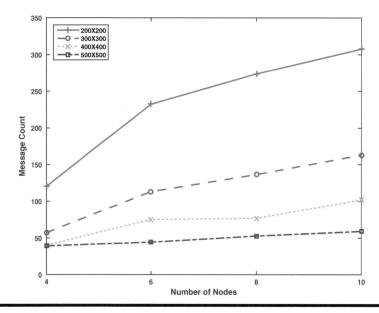

Figure 6.6 **The number of delivered messages with respect to the number of nodes and the size of the monitored application area. The number of participants is set to 1, while the number of nodes is varied between 4 and 10.**

We evaluate the accuracy and integrity of the collected data next. The number of nodes is set to 10 and the number of participants is set to 4. The size of the application area is 500 m × 500 m. We introduce the concept of malicious participants for the rest of the experiments. Malicious participants are assumed to always manipulate the data they collect from sensors and forward the altered data to the data center. In the following experiments, we considered a sample space with five possible values for the sensed data, as given in Table 6.3. Depending on the application, sample space may denote incremental or particular values, as discussed earlier. We have employed AADV and FDV approaches to resolve conflicting data and validate the data for corresponding nodes.

Success rates for the mentioned approaches are illustrated in Figures 6.7 and 6.8 in terms of accuracy and integrity, respectively. Integrity evaluates whether the precise data can be obtained. This metric is primarily useful when the data values are not incremental but particular. On the other hand, accuracy evaluates how close the obtained data is to the original data. For instance, let us assume that the original data is "very light." In terms of integrity, there is no difference whether the validated data is "light" or "heavy," and both will be marked as a failure. However, we assess the validated data based on its distance to the original value in accuracy.

Figure 6.7 denotes that both approaches perform worse when the malicious activity is increased. Both AADV and FDV provide almost 90% success in validating the data when the number of malicious participants is 1. FDV performs better

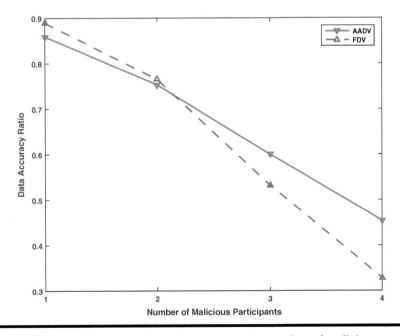

Figure 6.7   Data accuracy ratio with respect to the number of malicious participants when the number of participants is set to 4.

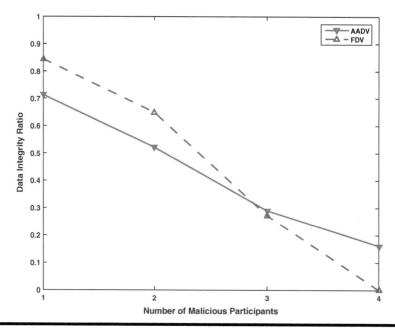

Figure 6.8   Data integrity ratio with respect to the number of malicious partici-pants when the number of participants is set to 4.

initially but is outperformed by AADV when the number of malicious participants is 3 or more. The decline in the performance of the FDV can be attributed to its working principle in validating the data. Recall that FDV considers the frequency of the collected data for validation. When the number of honest participants is more than or equal to the number of malicious participants, FDV performs bet-ter. However, when the number of malicious participants is increased further, fre-quency becomes misleading.

Figure 6.8 demonstrates the results in terms of integrity of the collected data when malicious participants exist in the network. Initially, FDV reaches a success rate of 85% while AADV provides 71% integrity. However, the performance of the FDV declines rapidly, and it is outperformed by AADV when the number of mali-cious participants is more than half of the total participants. If all the participants are malicious, the success rate drops to 0% for FDV. AADV, on the other hand, provides 16% integrity.

6.6 Conclusion and Future Issues

The availability of several sensors on ubiquitous devices such as smartphones, smart watches, and intelligent vehicles has enabled large-scale sensing tasks to be carried out by public crowds. On the one hand, this sensing model avoids the installation of custom hardware and its maintenance. On the other hand, the reliability of the application is highly dependent on the participants. Considering inherent human mobility, participants provide ample opportunities to improve some of the network performance metrics, such as connectivity and coverage. Participants may even be employed to provide connectivity to an otherwise disconnected network of sensor nodes and enable a CaNET. However, in the best case, nodes can be intermittently connected with the data center due to random mobility. Depending on the number of participants and their movement trajectories, some of the nodes may never be able to send their data to the data center. Therefore, the number of participants should be increased through incentive mechanisms. Another major challenge is the reliability of the participants. In the case of malicious participants, the sampled data can be altered. Consequently, the data center will likely receive conflicting data from different participants. In order to provide a certain level of reliability, data conflicts must be resolved and the actual data must be identified. Considering malicious participants, we defined two different metrics, accuracy and integrity, in order to assess the data quality. While the accuracy metric evaluates the disparity between the obtained value and the expected value, the integrity metric expects the exact value. Accuracy is useful when sensor data is incremental (e.g., brightness obtained from a light sensor). Integrity, on the other hand, can be employed when precise measurement is required (e.g., directions obtained from a gesture sensor). To resolve conflicting data, we present two novel approaches based on arithmetic average and frequency considering the defined metrics.

Data quality in CaNET can be assessed in terms of accuracy, integrity, and latency. In this chapter, we investigated accuracy and integrity metrics, and we are planning to consider latency in a future work. Latency is a major issue when participants employ only low-rate, short-range wireless communication means. In such a network, a BS must be present within the network to provide an Internet connection. In order to deliver sampled data to the data center, participants should visit the BS. This scheme introduces additional delay for the data delivery. Another issue that warrants additional investigation is the mobility model. Besides synthetic mobility models, which simulate human behavior, real traces can be obtained and considered in future studies.

References

1. Iphone 7 specifications. https://www.apple.com/iphone-8/specs. Accessed 2017-09-11.
2. Galaxy S8 specifications. http://www.samsung.com/global/galaxy/galaxy-s8/specs. Accessed 2017-09-11.
3. Garmin Vivoactive 3 specifications. https://buy.garmin.com/en-US/US/p/571520 #specs. Accessed 2017-09-11.
4. Tesla Model X specifications. https://www.tesla.com/sites/default/files/model_x_owners_manual_north_america_en.pdf. Accessed 2017-09-11.

5. Vandrico's wearable devices database. http://vandrico.com/wearables/wearable-technology-database. Accessed 2017-09-11.

6. Vandrico's wearable devices database. http://www.gartner.com/newsroom/id/3790965. Accessed 2017-09-11.

7. Liu, Y., Guo, B., Wang, Y., Wu, W., Yu, Z., & Zhang, D. (2016, September). Taskme: Multi-task allocation in mobile crowd sensing. In *Proceedings of the 2016 ACM International Joint Conference on Pervasive and Ubiquitous Computing* (pp. 403–414). New York: ACM.

8. Wang, L., Zhang, D., Pathak, A., Chen, C., Xiong, H., Yang, D., & Wang, Y. (2015, September). CCS-TA: Quality-guaranteed online task allocation in compressive crowdsensing. In *Proceedings of the 2015 ACM International Joint Conference on Pervasive and Ubiquitous Computing* (pp. 683–694). New York ACM.

9. Xiong, H., Zhang, D., Chen, G., Wang, L., Gauthier, V., & Barnes, L. E. (2016). iCrowd: Near-optimal task allocation for piggyback crowdsensing. *IEEE Transactions on Mobile Computing*, 15(8), 2010–2022.

10. Cheung, M. H., Southwell, R., Hou, F., & Huang, J. (2015, June). Distributed time-sensitive task selection in mobile crowdsensing. In *Proceedings of the 16th ACM International Symposium on Mobile Ad Hoc Networking and Computing* (pp. 157–166). New York ACM.

11. Antonić, A., Marjanović, M., Pripužić, K., & Žarko, I. P. (2016). A mobile crowd sensing ecosystem enabled by CUPUS: Cloud-based publish/subscribe middleware for the Internet of Things. *Future Generation Computer Systems*, 56, 607–622.

12. Cardone, G., Corradi, A., Foschini, L., & Ianniello, R. (2016). Participact: A large-scale crowdsensing platform. *IEEE Transactions on Emerging Topics in Computing*, 4(1), 21–32.

13. Ra, M. R., Liu, B., La Porta, T. F., & Govindan, R. (2012, June). Medusa: A programming framework for crowd-sensing applications. In *Proceedings of the 10th International Conference on Mobile Systems, Applications, and Services* (pp. 337–350). New York: ACM.

14. Long-term evolution. https://en.wikipedia.org/wiki/LTE_(telecommunication). Accessed 2017-09-11.

15. Ma, M., & Yang, Y. (2008, April). Data gathering in wireless sensor networks with mobile collectors. In *Parallel and Distributed Processing, 2008 (IPDPS 2008), IEEE International Symposium on* (pp. 1–9). New York: IEEE.

16. Senturk, I. F., Akkaya, K., Senel, F., & Younis, M. (2013, June). Connectivity restoration in disjoint wireless sensor networks using limited number of mobile relays. In *Communications (ICC), 2013 IEEE International Conference on* (pp. 1630–1634). New York: IEEE.

17. Janansefat, S., Akkaya, K., Senturk, I. F., & Gloff, M. (2013, October). Rethinking connectivity restoration in WSNs using feedback from a low-cost mobile sensor network testbed. In Local Computer *Networks Workshops (LCN Workshops), 2013 IEEE 38th Conference on* (pp. 108–115). New York: IEEE.

18. Wang, G., Cao, G., La Porta, T., & Zhang, W. (2005, March). Sensor relocation in mobile sensor networks. In *INFOCOM 2005: 24th Annual Joint Conference of the IEEE Computer and Communications Societies, Proceedings IEEE* (Vol. 4, pp. 2302–2312). New York: IEEE.

19. Ghosh, A., & Das, S. K. (2008). Coverage and connectivity issues in wireless sensor networks: A survey. *Pervasive and Mobile Computing*, 4(3), 303–334.

20. Wang, B., Lim, H. B., & Ma, D. (2009). A survey of movement strategies for improving network coverage in wireless sensor networks. *Computer Communications*, 32(13), 1427–1436.

21. Basagni, S., Carosi, A., Melachrinoudis, E., Petrioli, C., & Wang, Z. M. (2008). Controlled sink mobility for prolonging wireless sensor networks lifetime. *Wireless Networks*, 14(6), 831–858.

22. Wang, W., Srinivasan, V., & Chua, K. C. (2005, August). Using mobile relays to prolong the lifetime of wireless sensor networks. In *Proceedings of the 11th Annual International Conference on Mobile Computing and Networking* (pp. 270–283). New York: ACM.

23. Younis, M., Senturk, I. F., Akkaya, K., Lee, S., & Senel, F. (2014). Topology management techniques for tolerating node failures in wireless sensor networks: A survey. *Computer Networks*, 58, 254–283.

24. Ma, H., Zhao, D., & Yuan, P. (2014). Opportunities in mobile crowd sensing. *IEEE Communications Magazine*, 52(8), 29–35.

25. Low-rate wireless personal area networks. https://en.wikipedia.org/wiki/IEEE_802.15.4. Accessed 2017-09-11.

26. He, D., Chan, S., & Guizani, M. (2015). User privacy and data trustworthiness in mobile crowd sensing. *IEEE Wireless Communications*, 22(1), 28–34.

27. Zhang, X., Yang, Z., Sun, W., Liu, Y., Tang, S., Xing, K., & Mao, X. (2016). Incentives for mobile crowd sensing: A survey. *IEEE Communications Surveys & Tutorials*, 18(1), 54–67.

28. Jaimes, L. G., Vergara-Laurens, I. J., & Raij, A. (2015). A survey of incentive techniques for mobile crowd sensing. *IEEE Internet of Things Journal*, 2(5), 370–380.

29. Peng, D., Wu, F., & Chen, G. (2015, June). Pay as how well you do: A quality based incentive mechanism for crowdsensing. In *Proceedings of the 16th ACM International Symposium on Mobile Ad Hoc Networking and Computing* (pp. 177–186). New York: ACM.

30. Jin, H., Su, L., Chen, D., Nahrstedt, K., & Xu, J. (2015, June). Quality of information aware incentive mechanisms for mobile crowd sensing systems. In *Proceedings of the 16th ACM International Symposium on Mobile Ad Hoc Networking and Computing* (pp. 167–176). New York: ACM.

Chapter 7

Mobile Sensor and Cloud Computing with the Aid of the Crowd

Tian Wang, Jiyuan Zhou, Hao Luo, Yang Li, and Md Zakirul Alam Bhuiyan

Contents

7.1　Introduction

With the combination of wireless sensor networks (WSNs) and cloud computing, the sensor–cloud system (SCS) has arisen in recent years. This new system brings new challenges to traditional data collection, target localization, data storage, trust mechanisms, and so forth [1, 2]. For example, the efficiency of data collection in WSNs cannot fit the high processing abilities of the cloud, and the accuracy of localization cannot meet the need of actual applications. Fortunately, besides the normal sensors in the WSNs, there are some mobile elements, such as mobile phones, smart cars, and robots [3]. These mobile elements are called a "crowd," which can move and collaborate with each other. By utilizing the mobility and interoperability of the crowd, the above challenges can be solved effectively. Our main work includes two parts: data collection with the crowd and crowd-aided localization and tracking.

In realistic applications, data collection from WSNs to the cloud becomes a bottleneck because of the poor communication ability of WSNs. Since sensors have low bandwidth and low energy supplies, it is difficult to meet the requirements of real-time data delivery, especially in delay-sensitive applications. In order to improve the efficiency of data collection in WSNs, we introduce the mobile crowd, such as mobile sensor, mobile agent (MA), and mobile sink (MS), which can receive data transmitted from sensors directly or through fewer-hop wireless transmission. Due to the natural mobility and relatively strong storage ability, the mobile crowd is able to act as an intermediate relay in the proposed network architecture. Compared with traditional WSNs, WSNs with the aid of the mobile crowd have a longer lifespan and lower energy consumption, since the mobile crowd can undertake part of transmission work. In our work, we propose using mobile crowd help with data collection in sensor–cloud integration. A time-adaptive schedule algorithm (TASA) is designed to reduce data collection time and make the system sustainable.

Crowds can also be used for wireless localization. Wireless localization has attracted considerable research interest because of the increasing demand for location-based services in the military, industry, and WSNs. The Global Positioning System (GPS) is one of the most well-known and widely used localization techniques. However, two factors limit its application for mobile group users indoors. First, the localization error, which is approximately 10 m, is too large to satisfy the demand of location-based services. Second, it performs poorly in indoor environments, such as mines, markets, airports, and warehouses. Different from traditional methods, we present a crowd-aided localization. In the proposed approach, the mobile users (crowd) communicate not only with several fixed anchors but also with other mobile users. Then, the distances among them can be calculated. Based on this information, the mobile users can help themselves (as a whole) to get better localization performance.

7.2 Data Collection from WSNs to the Cloud with the Crowd

Data collection has always been a hot research topic in the WSNs field [4, 5]. With the surging of mobile crowd sensing, traditional sensing problems with the aid of the crowd have become a promising paradigm for cross-space and large-scale sensing. One of the most significant challenges WSNs face today is how to ensure the scalability of the network. For example, traditional sensors transmit data to the static sink through multiple-hop wireless communication [6, 7], which may result in energy depletion on sensors that close to static sinks. With the increasing number of disable sensors, network connectivity and coverage rate will accordingly decrease. Besides, real-time and low-cost sensing are also critical issues [8–10].

To address this problem, the notion of mobile elements has been proposed in recent years [11–13]. Sensors transmit data directly to mobile elements or through other sensors with fewer-hop wireless transmission; thus, the whole network can achieve lower energy consumption and a longer lifespan. Two types of mobile elements, MA and MS, are applied continually. MAs collect data from sensors and return to sinks [14–16]. For example, in [17], the MA is used to help with forest fire detection. It introduces a framework equipped with MAs to accelerate the detection rate of forest fire with the minimum consumption of energy. Compared with the MA, the MS can also move to collect sensed data but does not need to return to the station, because it is equipped with the ability of uploading data to the cloud. In [18], Arquam et al. proposed a routing algorithm with sink mobility in hierarchical WSNs to improve network lifetime by eliminating energy holes. They considered upper and lower bounds on delay while optimizing sojourn locations and sojourn time. In [19], Hou et al. designed an efficient-path algorithm VG-AFSA based on virtual grids to meet most applications' requirements for data latency.

The development of mobile crowd sensing sheds light on mobile sensor and cloud computing with the aid of crowds. Compared with traditional sensor networks, the key difference of mobile crowd sensing is the involvement of any amount of existing mobile devices for large-scale sensing. This feature offers mobile crowd sensing some advantages, like lower deployment cost and larger spatiotemporal coverage compared with static sensor network deployment. For example, mobile devices can be used to monitor the environment, like nature preservation, noise pollution measurement, and air pollution measurement. Moreover, in crowd-associated networks, the crowd can be viewed as an intermediate relay in the proposed network architecture. The crowd completes the communication gap between dedicated agents (sensors) and thereby improves the performance of the whole network.

In this chapter, we propose using mobile crowds as MSs helping with data collection in sensor–cloud integration. A TASA is designed to reduce data collection time and make the system sustainable. In TASA, the monitoring area is divided into several sectors equally and each MS is responsible for one sector. In each sector, parts of sensors are selected as polling points (PPs) that construct the trajectories of MSs and all the trajectories are nonoverlapping with each other. Besides, the minimum-cost spanning tree (MST) is adopted to design transmission routes for sensors to save energy consumption. Beyond that, we design two progressive schedule schemes to adjust data collection time and save energy. One focuses on reducing the MS's moving time, and the other aims to balance the load of each MS. The main contributions are summarized as follows:

1. Compared with the traditional data collection problem in WSNs, the problem of data collection in sensor–cloud integration is more complex. We propose a delay constraint problem caused by the limited transmission ability of WSNs in sensor–cloud integration.
2. We use multiple MSs to improve the sustainability of the sensor–cloud and design a time-adaptive algorithm aimed at collecting data from WSNs to the cloud within a specific time, with several provable properties. Beyond that, the energy consumption of sensors is optimized based on the property of MST.
3. We conduct extensive simulation experiments to evaluate the performance of the proposed algorithm, and the experimental results validate the effectiveness of our proposed algorithm.

7.2.1 Problem Description

In this chapter, we assume that the WSN consists of N sensors, denoted by a set $S = \{S_1, S_2, \ldots, S_N\}$. The set $K = \{MS_1, MS_2, \ldots, MS_M\}$ represents M MS$_s$ (MS$_s$). For any $S_i \in S$, the sensing data rate is C bytes per second, the single-hop latency is t seconds, and the communication radius is R meters. Any MS$_i \in K$ can receive data from sensors and upload it to the cloud. The throughput from sensors to the MS is D byte per second, and the uploading rate from the MS to the cloud is Q byte per second. The velocity of the MS is denoted by v meters per second. We focus on the problem of data collection from WSNs to the cloud (hereafter referred to as the DCWC problem). The goal of this chapter is to reduce the delivery time to a limited time. When the delivery time satisfies the requirement, the energy consumption will be optimized.

A simple example is illustrated in Figure 7.1, where four MSs are deployed to collect data from the WSN to the cloud. The circles and small cars stand for fixed sensors and MSs, respectively. The routes of MSs are shown as a dotted line.

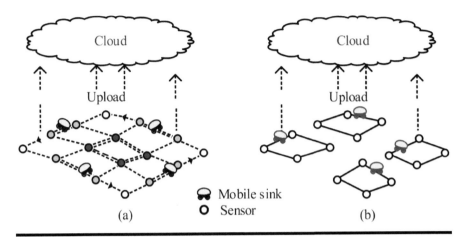

Figure 7.1 **An example of data collection from the WSN to the cloud with multiple MSs. (a) Four MSs are deployed to collect data from the WSN to the cloud. The routes of the MSs are random. The sensor's color represents the number of MSs visiting this sensor. The deeper the color is, the more MSs that are visiting this sensor. (b) The routes of MSs are adjusted and all the sensors are visited by only one MS.**

In Figure 7.1a, four routes have the overlapping parts, and the deeper the sensor's color, the more the sensor is visited by MSs. This phenomenon means that delivery time can be optimized. Then the route of each MS is adjusted in Figure 7.1b. As we can see, each MS is responsible for four sensors. What is more, each sensor is visited only once and the route length of each MS is reduced. Thus, it is essential to design a schedule algorithm to reduce data collection time in sensor–cloud integration.

We have the following theorem regarding the complexity of the DCWC problem:

Theorem 7.1: For the DCWC problem, the problem of designing the optimal algorithm is NP-hard.

Proof: We prove this theorem by showing a special case of DCWC, in which $M = 1$, the sensor transmission radius is zero, and uploading time is zero. In this case, the DCWC is equivalent to finding a shortest path visiting all the given sensors. Note that in order to minimize the path length, any optimal solution would not visit the same sensor twice; otherwise, we can make it shorter by using triangular inequality. Therefore, finding an optimal solution of this special case of DCWC is equivalent to finding an optimal solution of the Hamiltonian path problem (i.e., finding a path to visit all sensors with the minimum length), which is a well-known NP-hard problem.

We model the connectivity of sensors in the WSN as an unoriented weighted graph $G = \{V_{se}, E_{se}\}$, where $V_{se} = S, E \in \{V_{se} \times V_{se}\}$ is a set of edges, where $E_{i,j} \in E_{se}$ is the edge if the distance of $d_{i,j}$ between S_i and S_j is smaller than R. Then, the graph G is transformed to $MST = \{T_{node}, T_{edge}\}$, where $T_{node} = V_{se}$ and $T_{edge} \subseteq E_{se}$. Each MS will visit some selected sensors called polling points, denoted by $\{PP \subseteq S \mid P_1, P_2, \ldots, P_k\}$. Through visiting all the PPs, the sensory data in WSN would be collected and uploaded to the cloud. With multiple MSs synchronously working, the network delivery delay can be estimated as follows:

$$T_{net} = \text{Max}\left\{T_{MS_1}, T_{MS_2}, \ldots, T_{MS_M}\right\}, \tag{7.1}$$

where T_{MS_i} is the delivery time of T_{MS_i}, and the network delivery time is the max time among all MSs. As for each individual T_{MS_i}, it consists of four parts, formulated as follows:

$$T_t = \frac{\sum_{j=1}^{s} C}{D}, \tag{7.2}$$

$$T_d = \frac{\sum_{j=1}^{s} C}{Q}, \tag{7.3}$$

$$T_h = \sum_{j=1}^{s} h_j * t, \tag{7.4}$$

$$T_m = \frac{L_{tsp}}{v}, \tag{7.5}$$

T_t is the transmission time from the sensor source to the sink. T_d is the uploading time from the sink to the cloud. T_h is the multiple-hop delay, and T_m is the traveling time of the MS. In all arithmetic expressions, s is the number of sensors assigned to MS_i. The variable h_j is the amount of hops from sensor S_j to MS_j, and L_{tsp} is the length that MS_i traveled. We assume that each MS can upload data to the cloud at any time. Therefore, each T_{MS_i} can be calculated as Equation 7.6.

$$T_{MS_i} = \frac{\sum_{j=1}^{s} C}{D} + \sum_{j=1}^{s} h_j * t + \text{Max}\left(\frac{\sum_{j=1}^{s} C}{Q}, \frac{L_{tsp}}{v}\right), \tag{7.6}$$

Based on Equation 7.6, we can conclude that h_j, L_{tsp}, and s are the main external factors affecting T_{MS_i}, which motivates us to design the algorithm in the next section. Note that sensory data can be uploaded when MS is moving according to the existing routing algorithm [20].

7.2.2 Time-Adaptive Schedule Algorithm

The design of the algorithm can be divided into three subproblems: (1) how to distribute the task to each MS reasonably; (2) how to design the delivery time-adaptive mechanism, and (3) how to reduce the energy consumption when the delivery time meets the requirement. In order to address these issues, the solution has three corresponding subalgorithms. In first subalgorithm, called the partition and delivery design algorithm (PDDA), the monitoring area is partitioned off M sectors equally and each MS is responsible for one sector. In each sector, an MST is constructed and the edges of the tree are the delivery routes of sensors. As for the second subalgorithm, called the polling point selection algorithm (PPSA), some sensors are selected to serve as PPs in each sector. The PPs in the same sector constitute the trajectory of the MS. In the third algorithm, called the time schedule algorithm (TSA), the number of PPs is adjusted based on the delivery time and the number of sensor in each sector will be adjusted to be balanced.

Figure 7.2 shows the main process of TASA. The rectangle area represents the coverage area, and the dotted black circle is the circumscribed circle of the rectangle in Figure 7.2a. The area is divided into three sectors equally, denoted by $M1$, $M2$, and $M3$. Three MSTs are generated in each sector. The edges of MSTs are delivery paths of sensors. In Figure 7.2b, parts of sensors are selected as PPs represented by the gray star nodes. The black dotted lines among stars are the trajectories of MSs. The dotted big arrows mean that sensors deliver data to MSs when MSs rest at PPs. When delivery time cannot satisfy the requirement, the

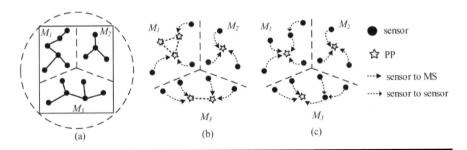

Figure 7.2 The basic idea of TASA. (a) It shows partition in the monitoring area, where the dotted circle is the circumcircle of the rectangle and the dotted lines in the rectangle represent the boundary. The boundary is designed based on the degree in the circumcircle. (b) Some sensors are selected as PPs, and all PPs construct the trajectory of the MS, like the dotted line among the stars. (c) Some sensors are no longer PPs and the trajectory of the MS is redesigned.

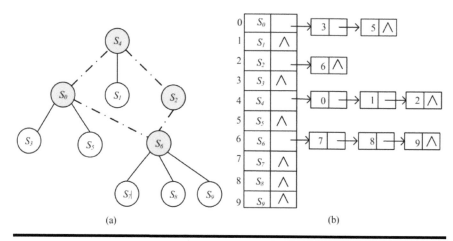

(a)   (b)

Figure 7.3   The strategy of PP selection. (a) In MST, the sensor that connects with more than one other sensor is selected as the PP, like the gray nodes. (b) This is a physical structure showing a clear relationship among sensors.

trajectory of MS will be adjusted by reducing the number of PPs. In Figure 7.3c, some sensors no longer serve as PPs and deliver data to the MS via multiple hops, where the

Algorithm 7.1 Partition and Delivery Design Algorithm

Input: Each sensor's location $(S_i.x; S_i.y)$, the number of sensors is N. Two sets $T_{node} = T_{edge} = \Phi$ represent the nodes and the edge of MST, respectively. The V_{sec} is a set of sensors in same sector;

Output: M graphs $G(V_G, E_G)$ and MSTs, denoted as $T(V_T, E_T)$;

1: **while** i, j are smaller than N **do**

2:     $E_{i,j} = \sqrt[2]{\left(S_i.x - S_j.x\right)^2 + \left(S_i.y - S_j.y\right)^2}$;

3:     **if** $E_{i,j} > R$ **then**

4:         $E_{i,j} = \infty$;

5:     **end if**

6: **end while**

7: **while** $T_{node} \uparrow V_{sec}$ /*Prim Algorithm*/**do**

8:     find $MinE_{u,v}, u \in T_{node}, v \in S$;

9:     **if** $v \notin T_{node}$ **then**

10:         $T_{node} = T_{node} \cup v$;

11:         $T_{edge} = T_{edge} \cup E_{u,v}$;

12:     **end if**

13: **end while**

dotted small arrows mean data transmission between sensors. As only one PP is left in each sector, the MS stays at this PP. All sensors send their data to the MS via multiple-hop transmission. Considering the different number of sensors in each sector, we design a balanced strategy to adjust the task allocated in each sector.

This subalgorithm is the first part of TASA, which focuses on network partition and delivery route design. In order to simplify partition, the monitoring area is partitioned based on degree. More specifically, we use a min rectangle to include all sensors. However, it is difficult to partition the rectangle off the M region with the same perimeter. So, we partition the coverage area based on its circumscribed circle. Coverage area is equally divided into M sectors with a central angle $2\pi/M$. Second, the connectivity among sensors in each sector constitutes M weighted graphs $\{G_{MS_i} \mid i = 1, 2, \ldots, M\}$. Besides, the weight of the edge can be calculated by Equation 7.7.

$$E_{i,j} = \sqrt[2]{\left(S_i.x - S_j.x\right)^2 + \left(S_i.y - S_j.y\right)^2}, \tag{7.7}$$

$S_i.x$, $S_i.y$ represent the abscissa and ordinate of the sensor, respectively. If the value of edge $E_{i,j}$ is larger than radius R, the weight of this edge would be set as infinity. Then based on the Prim algorithm, M weighted graphs are transformed to M MSTs, denoted as $\{Mst_i \mid i = 1, 2, \ldots, M\}$. According to the properties of MST, when delivery time T_{MS_i} is smaller than the latency requirement T_{spe}, the transmission consumption of sensors is the minimum.

The PPs make it possible for MS to visit parts of sensors collecting all sensory data. A basic election principle can be described as follows. When S_j is within the transmission range of S_i, MS can stay at S_i to collect both S_i and S_j. Consequently, MS can visit part of the sensors to complete all data collection.

A reasonable selection of PP can reduce the moving time of MS. In the PDDA, the sensors that construct an MST can be classified into three types: root node, potential PP node, and leaf node. The root

Algorithm 7.2 Polling Point Selection Algorithm

Input: The MST $T = (T_{node}, T_{edge})$; root node: u; V_i represents the set of sensors in sector i; V_{is} represents the set of sensors that are visited;
Output: The set of PP, denoted by $\Xi = P1, P2, \ldots, Pk$, and the visiting order of PPs, denoted by Λ.
1: $V_{is} = u$;
2: **while** $V_{is} \neq V_i$ **do**
3:    **if** $E_{u,v} \in T_{edge}$ and $v \neq V_{is}$ **then**
4:       $V_{is} = \{V_{is} \cup v\}$;
5:       **if** sensor v connects with more than one sensor **then**
6:          $\Xi = \Xi \cup v$; /*sensor v is selected as PP*/
7:       **end if**
8:    **end if**
9: **end while**

10: **while** the set $\Lambda \neq \Xi$ **do**
11:     next $PP= PP = \text{Min}\left(\text{emph } S_u \rightarrow S_w\right)$;/*find the next minimum path*/
12:     $\Lambda = \{\Lambda \cup S_w\}$;
13:     updating the start node as S_w;
14: **end while**

node is the starting point of the MS. The potential PP node is a kind of sensor that is connected with more than one sensor directly. The leaf node is a sensor that is connected with only one sensor. In Figure 7.3a, it is easily known that the MS can visit S_6, S_2, S_0, and S_4 to gather all sensory data; then these four sensors are selected as PPs, like the gray nodes. The dotted line is the trajectory of the MS. On the one hand, the trajectory of the MS can be optimized by a greedy algorithm. On the other hand, when the MS rests at the PP, the leaf nodes deliver data to the MS through a single hop to realize the minimum energy consumption. Figure 7.3b presents a physical storage structure of the MST called a children-linked list. It shows a clear relationship among sensors in the MST. S_4 is the root node. S_6, S_0, S_2 are the potential PP nodes. S_1, S_3, S_5, S_7, S_8, S_9 are the leaf nodes.

In two stages, we have designed an initial trajectory for each MS. It is well known that the employment of MSs can balance the load of sensors and prolong network lifetime. However, due to the limited speed of MSs, a lot of time is always wasted for moving. To solve this problem, two kinds of schedule schemes are designed to adjust the delivery time until the delay requirements are satisfied.

Algorithm 7.3 Time Schedule Algorithm

Input: The time requirement T_{spe}; the delivery time of each sector, denoted by $\{T_{MS_1}, T_{MS_2}, \ldots, T_{MS_M}\}$; the set of PP, $\Xi = \{PP_1, PP_2, \ldots, PP_k\}$
Output: The delivery route of each sensor and the trajectory of each MS
1: **while** $i \leq M$ **do**
2:     **while** $T_{MS_i} > T_{spe}$ **do**

3:         $PP_i = \text{Min degree}\{PP_1, PP_2, \ldots, PP_k\}$;

4:         $\Xi = \Xi - PP_i$; /* the PP_i; is no longer polling point */
5:         the trajectory of MS is reprojected;
6:     **end while**
7: **end while**
8: the sectors are divided into two sets, named A and B;
9: **if** $T_{MS_i} > T_{spe}$ **then**
10:     put sector i into set A
11: **else** Put sector i into set B
12: **end if**
13: **while** $A \neq \varnothing$ and $B \neq \varnothing$ **do**
14:     Selecting sector $SS = \text{Max} T_{MS_x}\{A\}$ and $ES = \text{Min} T_{MS_y}\{B\}$;

15:    sensor $\{S_\xi \in SS | S_\xi \overset{\min}{\to} ES\}$ changes its route and delivers data to the *MS* responsible for *ES*
16:    sets A and B
17: **end while**

Schedule scheme 7.1. In the initial phase, all sensors deliver data to the MS via a single hop, like the white nodes in Figure 7.4a. The gray nodes are PPs, and the dotted lines are trajectories of the MS. Then we calculate collection time T_{MS_i} by Equation 7.6. If T_{MS_i} is greater than T_{tpe}, some PPs will be removed from the set of PP and added into the general sensor set. The PP that is connected with fewer sensors is chosen first, such as S_4 in Figure 7.4b. Its child node S_2 sends data to MS via S_4 with two hops. Moreover, the trajectory of MS will be redrawn after a round of selection. Until the delivery time meets the requirement, the trajectory of MS is fixed. When only one PP is left in the sector, the moving time of MS is zero, and the delivery time is the sum of multiple-hop delay and uploading time. As shown in Figure 7.4c, the dotted arrows represent the order that PP becomes a general sensor.

Schedule scheme 7.2. Different from schedule scheme 7.1, we consider the factor that the number of sensors is different in each sector. When the latency demand T_{tsp} is extremely small, parts of sectors may not satisfy the time requirement. Moreover, the network delivery time is the time when the last sensor is collected, as shown in Equation 7.1. Consequently, this schedule algorithm aims to balance the task of each MS, which can be described in the following four steps.

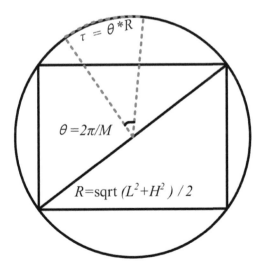

$\tau = \theta * R$

$\theta = 2\pi / M$

$R = sqrt (L^2 + H^2) / 2$

R: radius θ: degree τ: arc length

Figure 7.4   The coverage area is divided into M sectors.

1. The sectors are classified into two sets: A and B, where set A includes the sector whose delivery time is smaller than T_{spe}. On the contrary, set B includes the sector whose delivery time is larger than T_{spe}.
2. In set B, the sector with max T_{MS_i} is selected as the start sector (SS). In set A, the sector with minimum T_{MS_i} is selected as the end sector (ES).
3. The sensor from SS that is closest to the ES delivers its data to the MS in the ES. In addition, the delivery times of both the SS and ES are updated. Then step 1 is repeated.
4. Iterate steps 2 and 3 until set A or set B is null.

7.2.3 Performance Analysis

We have proved several properties of the proposed algorithm. The notations that will be used in this part are summarized in Table 7.1.

Theorem 7.2: In the worst case, the travel distance of each MS satisfies this inequation: $L_{tsp} \leq \left(1 + \dfrac{\pi}{M}\right) \times \sqrt{L^2 + H^2}$

Proof: Assume that the sensor coverage area is a rectangle $L * H$ and its circumscribed circle is shown in Figure 7.4. The radius of the circle is $R = \sqrt{L^2 + H^2} / 2$. The circle is divided into M sectors equally based on degree, and the central angle of each sector is $\theta = 2 * \pi = M$. The arc of each sector is $\tau = \theta * R$. The perimeter of each sector is $Dist_L = 2 * R + \tau$. In the worst case, MSs have to visit every sensor and the sensors are distributed at the margin of the sector. Consequently, the max travel distance of MSs is the perimeter of the sector, namely, $L_{tsp} \leq \left(1 + \dfrac{\pi}{M}\right) \times \sqrt{L^2 + H^2}$.

This property gives the moving distance of MSs an upper bound.

We have proved several properties of the proposed algorithm. The notations that will be used in this part are summarized in Table 7.1.

Theorem 7.3: In the worst case, the time complexity of TASA is $O(n^3)$, where n is the number of sensors.

Proof: In the first subalgorithm PDDA, the coverage area is divided into M sectors and its time complexity is $O(1)$. Then, based on the Prim algorithm, MST is generated in M sectors, and its time complexity is $M * O(n^3)$. In the general case, M is smaller than n. Now, we set M equal to n, so the time complexity is $O(n^3)$. In the second subalgorithm, PPSA, the time complexity of electing PPs is $O(n)$, but the time complexity of designing the trajectory of the MS based on the

Table 7.1   List of Notations

Notation	Meaning
M	Number of MSs
n	Number of sensors in WSN
R	Transmission radius of the sensor
L_{tsp}	Max distance that MS traveled
θ	Central angle of each sector
τ	Arc of each sector
$Dist_L$	Perimeter of each sector
T_{spe}	Time requirement
T_{net}	Delivery time of the network
$\{\mu_1, \mu_2, \ldots, \mu_M\}$	μ_i represents the set of sensors in δi
$\{\delta_1, \delta_2, \ldots, \delta_M\}$	$\ddot{E}$ represents the *ith* sector
$\{\vartheta_1, \vartheta_2, \ldots, \vartheta_M\}$	ϑ_i represents the set of PP in sector δ_i
$\{\gamma_1, \gamma_2 \ldots, \gamma_M\}$	γ_i represents sensors in sector ϑ_i
T_{ideal}	Theoretical optimum delivery time
$\bar{e}$	Energy consumption for unit length
$edge_{(i,j)}$	Weight between S_i and S_j
T_{init}	Delivery time in PPSA
E_{init}	Energy consumption in PPSA
E_i	Energy consumption of δ_i
$\ddot{T}$	Actual delivery time
$\ddot{E}$	Actual energy consumption

traveling salesman problem (TSP). TSP is $O(n^3)$ in the worst case. The third sub-algorithm, TSA, consists of two parts. In the first part, assuming that the sensor sets $\{u_1, u_2, ..., u_M\}$ are all PPs, a PP becomes the general sensor each round. And the work will be repeated for M times because of M sectors, so the complexity of this part can be calculated by $\sum_1^M u_i$, which equals n, namely, $O(n)$. For the second part, assuming that one sensor changes its transmission route each time, and all the sensors will do this step, the complexity is $O(n)$. Therefore, the time complexity of TASA is $O(n^3)$ in the worst case.

Theorem 7.4: In the general value T_{spe}, TASA can realize the adaptive delivery time via schedule scheme 7.1, and if the time requirement T_{ideal} almost equals T_{spe}, the delivery time can be optimized by schedule scheme 7.2 so that each MS is responsible for the same number of sensors. For the time requirement T_{spe}, which is smaller than T_{ideal}, the delivery time is nearly the minimum in each sector.

Proof: Assume that sensors in the monitoring area are denoted by $S = \{\mu_1 \cup \mu_2 \cup ... \cup \mu_M\}$, where u_i represents a set of sensors in sector δ_i. θ_i is a set of the PPs in sector δ_i, and γ_i is a set of general sensors in sector δ_i. The relationship among μ_i, θ_i, γ_i is denoted by $\mu_i = \theta_i + \gamma_i$. Based on these notations, the delay model in Equation 7.6 can be simplified further. At the beginning of the TSA, all sensors deliver data to the MS via a single hop, so $h_i = (j = 1, 2..., N)$ equals 1 and the sum of $\sum_{j=1}^s C$ equals $\gamma * C$. L_{tsp} is the travel distance of visiting all PPs in θ_i, and we assume that MS can upload data to the cloud when the MS is moving. Therefore, the delivery time of the MS can be simplified as Equation 7.8:

$$T_{MS_i} = \frac{\gamma_i * C}{D} + \gamma_i * t + \text{Max}\left(\frac{L_{tsp}}{V}, \frac{\gamma_i * C}{Q}\right) \tag{7.8}$$

As the delivery time cannot meet the application requirement, the TSA would adjust the path of the MS to reduce moving time. For the first case, the time constraints T_{spe} are smaller than T_{ideal}. Until T_{MS_i} is smaller than T_{spe}, parts of PPs are transformed to general sensors and both sets γ_i and θ_i are not null. In this situation, T_{MS_i} is smaller than T_{spe} and h_j equals always 1. Furthermore, when T_{spe} nearly equals T_{ideal}, the goal of the problem can be regarded as minimizing the delivery time. Therefore, the PP would be fixed in each sector and most of the sensors deliver data to the MS via multiple hops, such as Figure 7.5c, where $\theta_i = 1$; $\mu_i = \gamma_i + 1$. One thing to note is that T_{tsp} equals zero; then the delivery time can be shown as in Equation 7.9.

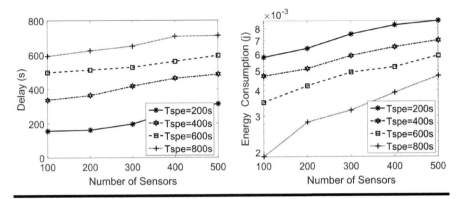

Figure 7.5 (a) Number of sensors vs. delay. (b) Number of sensors vs. energy consumption.

$$T_{MS_i} = \frac{\gamma_i * C}{D} + \frac{\gamma_i * C}{Q} + \sum_{j=1}^{\gamma_i} b_j * t \qquad (7.9)$$

According to the analysis of Equation 7.9, it is obvious that γ_i is the key factor influencing delivery rime. Through the schedule scheme 7.2 in the TSA, the final status is that all the MSs are responsible for the same number of sensors, denoted as $\gamma_1 = \gamma_2 = \ldots = \gamma_M$. Therefore, if T_{spe} nearly equals T_{ideal} or T_{spe} is much smaller than T_{ideal}, Theorem 7.4 is validated, denoted as $T_{MS_i} \approx T_{ideal}$.

Theorem 7.5: If the solution for designing the trajectory of the MS is the optimal in the PPSA, the sensor energy consumption is the minimum.

Proof: The initial delivery time and network energy consumption can be calculated by Equations 7.10 and 7.11:

$$T_{init} = \frac{\sum_{j=1}^{s} C}{D} + \sum_{j=1}^{s} b_j * t + \text{Max}\left(\frac{\sum_{j=1}^{s} C}{Q}, \frac{L_{tsp}}{v} \right) \qquad (7.10)$$

$$E_{init} = \overline{e} * \sum_{i=1}^{M} \sum T_{edge} \qquad (7.11)$$

Based on the property of the MST, the sum length of edges is the minimum in each sector; namely, ΣT_{edge} is minimum. Therefore, at the initial stage, the energy consumption E_{init} is the minimum in each sector. Now giving a PP set $\theta_k = (PP_1, PP_2, \ldots, PP_r)$, the child nodes of PP_i are denoted as $\{S_a, S_b, \ldots, S_k\}$.

Δt is the reduced time and ΔE is the increased energy consumption via the TSA. The travel distance of MS will decrease, denoted by $\Delta L = L_{\text{tsp}} - \sum_{j=i-1}^{i+1} d_{i,j} \, (i \neq j)$, where $d_{i,j}$ is the distance between PP_i and PP_j. The hop will increase, denoted as $\Delta h = \sum h_w (w = S_a, S_b, \ldots, S_k)$. The transmission distance of sensors will increase, denoted as $\bar{L} = \sum_{i=1}^{\Delta h} l_i$, where the notation l_i is the length of each hop. Consequently, the decreased delivery time can be denoted as $\Delta T = \dfrac{\Delta L}{v} - \Delta h * t$. The increased energy consumption can be denoted as $\Delta E = \bar{e} * \Sigma_{i=1}^{\Delta h * t}$. The relationship between actual energy consumption $\ddot{E}$ and actual delivery time $\ddot{T}$ satisfies Equations 7.12 and 7.13:

$$\ddot{T} = T_{\text{init}} - \frac{L_{\text{tsp}} - \sum_{j=i-1}^{i+1} d_{i,j}}{v} + \sum h_w * t, \quad w = S_a, S_b, \ldots, S_k \tag{7.12}$$

$$\ddot{E} = E_{\text{init}} + \bar{e} * \sum_{i=1}^{\Delta h * t}, w = S_a, S_b, \ldots, S_k \tag{7.13}$$

Based on Theorem 7.4, the delivery time $\ddot{T}$ can be optimized. When the solution for designing the trajectory of the MS in the PPSA is optimal, the value of Σh_w is optimal. According to the property of the MST, the value ΣL_i is the minimum and the value of ΔE is the minimum in each sector. Therefore, the value of $\ddot{E}$ is optimal with the minimum value of E_{init}.

7.2.4 Evaluations

To validate the effectiveness of our proposed algorithm, we conducted extensive simulations using MATLAB™ 2015a. We built a WSN consisting of 100 sensors deployed in a 100 m × 100 m rectangle region. The data-generating rate of each sensor is 5 bytes/s The transmission range of each sensor is 3 m, and its initial energy capacity is 30 J. The energy-consuming rate for the transmitting is 6×10^{-7} (j/bit), and for receiving rate it is 3×10^{-7} (j/bit). The speed of the MS is 3 (m/s). The data uploading rate from the MS to the cloud is 50 (byte/s). In the general case, there are five MSs deployed in the WSNs.

As part of the evaluation, two existing algorithms are also considered for comparison. The first is EMMS [21]. In this algorithm, multiple sinks were controlled to visit all the sensors and the tour of each MS is closed. The second is SG-MIP [22]. This algorithm iteratively partitions the monitoring area and selects the near-optimal route for MSs. In the simulations, these three algorithms are compared based on four metrics: the delivery delay, the energy consumption, the distances that the MS traveled, and the lifetime of the WNSs. The delivery delay is the max delivery time among all MSs. The energy consumption represents the max energy consumption among all sensors. The energy consumption of each sensor is calculated based on the energy model in [23].

Figure 7.5 demonstrates the delivery delay and energy consumption under the scenarios with different numbers of sensors. As shown in Figure 7.5a, when the number of sensor increases from 100 to 500, the delivery delay shows a rising trend. Due to the stringent deadline, when T_{spe} is smaller than 400 s, the delay requirement cannot be guaranteed. As T_{spe} increases, our algorithm performs well with different numbers of sensors, which is well consistent with the theoretical analysis (Theorem 7.4). Figure 7.5b shows the energy consumption for data collection. No matter what the value of the latency requirement is, as the number of sensors increases, the energy consumption decreases gradually since more sensors can deliver data to the MS through fewer hops. Moreover, the lower the value of the latency requirement is, the higher the energy consumption is. This is because the number of PPs will be cut down to shorten the travel time of the MS, which means more sensors will deliver data via multiple hops.

Figure 7.6 shows the delivery delay and energy consumption under scenarios with different numbers of MSs. In the Figure 7.6a, when the number of MSs increases, the delays with different demands all decrease. It is worth noting that the delay value has a nearly linear variation as the number of MSs is from 10 to 25 and T_{spe} is 800 s. This phenomenon reflects the step in the PPSA. When all the sensors send data to the MS through a single hop and the delivery time is smaller than T_{spe}, the collection time from the sensors to the MSs is constant. And the difference is more MSs uploading data to the cloud. As shown in Figure 7.6b, the energy

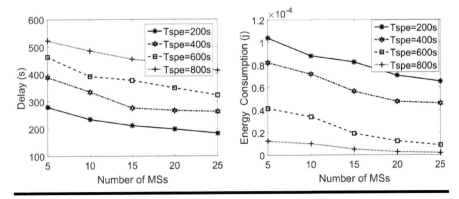

Figure 7.6 (a) **Number of MSs vs. delay.** (b) **Number of MSs vs. energy consumption.**

consumption decreases as the MSs increase. This is because more MSs are deployed to balance the load of each sensor. Besides, more sensors send data to MSs directly.

We now evaluate the algorithms on delay. As shown in Figure 7.7a, with the increase of the number of sensors, our proposed TASA achieves the best performance among these three algorithms. Specifically, the delay achieved by SG-MIP is about twice that of our algorithm. When the number of sensors is 250, the delay generated by EMMS is extremely high. In Figure 7.7b, when the number of MSs increases, all algorithms perform better, but TASA performs the best.

After comparing the effect of TASA with another two algorithms, EMMS and SG-MIP, we find two reasons for why our algorithm performs better. First, the radius of the sensor is a key element affecting the performance of the algorithm. As shown in Figure 7.8a, it is evident that the delay decreases with a bigger radius of sensor for all algorithms. Obviously, the influence of the radius on EMMS is the most apparent. It testifies that TASA has a better compatibility. For further clarity, we present the distances MSs traveled in all algorithms in Figure 7.8b. It is seen that TASA has the shortest distance and smallest difference among these three algorithms. This is because TASA can adjust the trajectory to fit the latency requirement. We now evaluate the effectiveness of the algorithm on energy consumption and network lifetime. In Figure 7.9a, the energy consumption achieved by EMMS stays steady due to the single-hop transmission of all the sensors. Beyond that, the energy consumption in TASA and SG-MIP decreases when the number of MS is added, and TASA realizes the lower energy consumption. Figure 7.9b shows the trend of network lifetime with the variance of numbers of MS_s. As more MSs are employed in WSNs, the lifetime becomes longer. This is because sensors have more opportunities to communicate with the MS directly. In contrast, the lifetime achieved by SG-MIP is shorter than that in our algorithm.

In Figure 7.10, we present the comparison result between TASA and MMSA, which was proposed in [24]. Figure 7.10a presents the comparison of delay with the increased number of MSs. Figure 7.10b implies the tendency of network lifetime

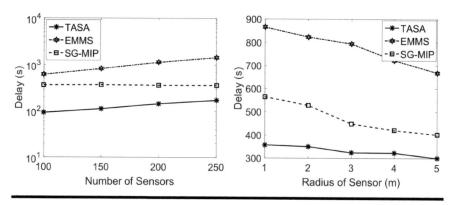

Figure 7.7 (a) Number of sensors vs. delay. (b) Number of MSs vs. delay.

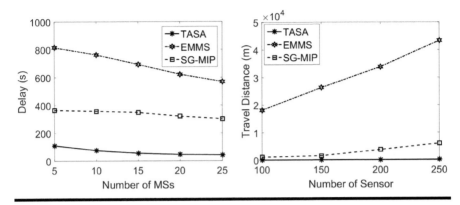

Figure 7.8 **(a) Radius of sensors vs. delay. (b) Number of sensors vs. travel distance of MS.**

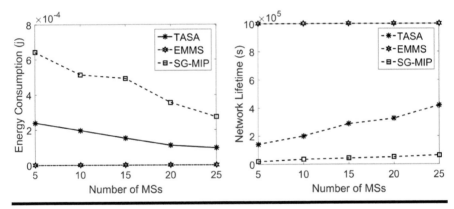

Figure 7.9 **(a) Number of MSs vs. energy consumption. (b) Number of MSs vs. network lifetime.**

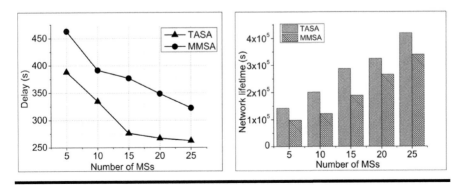

Figure 7.10 **(a) Number of MSs vs. delay. (b) Number of sensors vs. network lifetime.**

with the increased number of MSs. Both results obviously show that the improved TASA performs better than MMSA.

7.3 Crowd-Aided Interoperable Localization

In this section, we introduce our work about crowd-aided localization. As we mentioned, wireless localization has attracted considerable research interest because of the increasing demand for location-based services [25–27]. For example, indoor localization, which attempts to find the accurate positions of person and objects inside a building [28], can provide position information to provide services in various categories, including the location detection of firefighters in a building on fire, the location detection of products stored in a warehouse, the location detection of medical personnel or equipment in a hospital, finding tagged maintenance tools and equipment scattered all over a plant, tracking, monitoring, healthcare, and billing. However, for various reasons, such as the mobility of users, the instability of the wireless signal [29], multipath propagation, and non-line-of-sight (NLOS) paths [30], traditional localization techniques do not perform well for mobile users. Moreover, the solutions provided by these localization algorithms are acceptable only when there is sufficient distance information from the anchors. In certain scenarios, due to sparse anchor deployment, short radio communication ranges, and physical obstacles, a sufficient number of accurate measurements may not be available. In such cases, the localization accuracy declines sharply [31, 32]. Hence, developing a reliable algorithm to address these localization problems is important and urgent.

With the development of electronics manufacturing technology, mobility has been available for many devices, such as sensors [33, 34]; therefore, mobility-based research is more and more popular nowadays [35]. Mobility-aided localization methods, in which mobile anchors are exploited to aid in localization, have been investigated recently in studies such as [36] and [37]. A mobile anchor can follow mobile users in mobility-aided localization methods, which can improve the localization performance. However, it is well known that a moving anchor can cover only a small area within a reasonable time [38]. Obviously, adding more mobile anchors in the network area leads to better performance, but this incurs high costs. Moreover, the movement of the anchor increases energy consumption and introduces greater hardware support requirements, such as mobile devices. Therefore, mobility-aided methods still cannot solve the localization problem for mobile group users, which may include many users.

Another research interest is cooperative localization, in which the information among the users is exploited to help the localization [39, 40]. The most common idea is that some users are localized first, and then they can be exploited as virtual anchors to localize others who cannot be. Thus, there is a delay of positioning for the users, which also influences the localization accuracy. Another problem is that they focus on localizing the users who are immobile. If the users are mobile, due to the users' mobility and signal noise, the localization accuracy is limited. For example, in [41], the localization accuracy achieved in the static network seems feasible, but fails in dynamic networks. Patwari et al. present cooperative measurement-based statistical models and give the localization error analytically in [42]. They propose a distributed localization algorithm by successive refinement, which still cannot avoid the delay problem. In [43], the authors propose convex semidefinite programming (SDP) estimators specifically for the received signal strength (RSS)–based localization in both noncooperative and cooperative schemes, and the maximum likelihood estimator is appended to the convex estimator. However, the localization error is more than 1.5 m in most cases, which is too large.

Thus, we present a novel approach for localizing mobile group users with the crowd. In the proposed approach, the mobile users communicate not only with several fixed anchors but also with other mobile users (crowd). Then, the distances among them can be calculated. Based on this information, the mobile users can help themselves (as a whole) to get better localization performance. This localization idea is designed and combined with the extension of the Kalman filter, which further improves the localization performance. The main contributions of our work are listed as follows:

1. To localize mobile group users, we introduce the idea of interoperable localization, in which mobile users serve as the crowd to help locate each other.
2. A series of theoretical analyses regarding why the localization performance can be improved is presented.
3. We extend the Kalman filter algorithm to alleviate the influence of the noise in the environment and the instability of wireless signals.
4. We conduct extensive experiments to compare the proposed solution with traditional solutions, and the effectiveness is validated by the experimental results.

7.3.1 Crowd-Aided Localization Method

This section presents our proposed crowd-aided localization method. Different from other localization methods, the users in our proposed method can communicate with others and act as the crowd for each other, which can increase the reference information for localization. Figure 7.11 gives a simple example in which A_i ($i = 1, 2, 3, 4$) are the fixed anchors and U_i ($i = 1, 2, 3$) are then unlocalized mobile users. As shown in the figure, U_2 can receive signals from more than three anchors (A_1, A_2, A_3, and A_4). Obviously, its position can be estimated using a traditional fixed-anchor-based localization method. However, some users may not be able to successfully receive signals from enough anchors. For example, U_1 and U_3 may only receive signals from {A_1, A_2} and {A_3, A_4}, respectively. Therefore, they cannot be localized based on the traditional localization method.

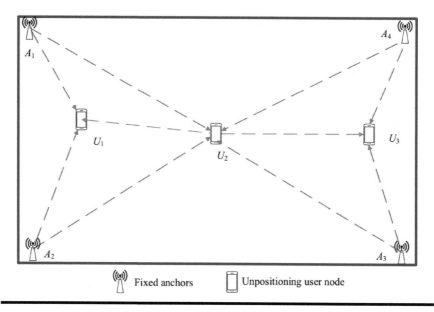

Figure 7.11 A simple illustration of mobile localization (U_1 and U_3 cannot be localized based on the traditional method with fixed anchors).

However, they can both receive signals from U_2, whose position has been estimated already, which motivates us to use U_2 as a mobile anchor to help U_1 and U_3 for localization.

The basic idea is that the mobile users communicate with the crowd and calculate the distances among them. Then, the users whose locations have already been localized can be exploited to localize the users whose locations have not. For one thing, the introduction of crowds can help to improve the localization probability. For another, this method is also expected to improve the localization accuracy, which will be discussed in the following section.

7.3.2 Performance Analysis of the Proposed Localization Method

In this section, we provide analyses to prove the effectiveness of the proposed localization method. Here, we consider the scenario in which there are three noncollinear anchors (A_1, A_2, and A_3) and three users (U_1, U_2, and U_3). We say the U_i can hear A_j if U_i can receive the anchor information from A_j. Assume that the distance from U_i to A_j is di, the distance from Ui to $A_{(j+1)}$ is d_j, the distance between U_i and U_j is d_{ij}, and the distance from A_i to A_j is $d_{A_iA_j}$. For ease of illustration, the analyses are based on 12 scenarios. Note that, based on this, the analyses can be easily extended to scenarios with more anchors and users. Table 7.2 shows a summary of the results, and the details are presented as follows.

Theorem 7.6: If each of the users can receive the signal of one and only one anchor (case 7 in Table 7.2), then the users can be localized.

Proof: Assume U_1 can hear A_1, U_2 can hear A_2, and U_3 can hear A_3, Then,

1. If $d_{A1A2} = d_{12} + d_1 + d_2$ or $d_{A1A2} = d_{12} - d_1 - d_2$, namely, U_1 and U_2 are on line A_1A_2, then U_1 and U_2 can be positioned. U_3 can also be localized as it can hear U_1, U_2, and A_3. In all, the users can be localized in this situation.
2. If one of U_1 and U_2 is on line A_1A_2, suppose U_1 here, then U_1 can be positioned. U_2 and U_3 can hear $\{U_1, A_2\}$ and $\{U_1, A_3\}$, respectively. Draw two circles with center $U1$ and radiuses d_{12} and d_{13}, and then their intersections

Table 7.2 Summaries of the Analyses

		A_1	A_2	A_3	Result			A_1	A_2	A_3	Result
1	U_1	★	★	★	×	7	U_1	★	☆	☆	√
	U_2	☆	☆	☆			U_2	☆	★	☆	
	U_3	☆	☆	☆			U_3	☆	☆	★	
2	U_1	★	★	☆	×	8	U_1	★	★	☆	√
	U_2	☆	☆	★			U_2	★	☆	☆	
	U_3	☆	☆	☆			U_3	☆	☆	★	
3	U_1	★	★		×	9	U_1	★	★	☆	√
	U_2	☆	☆	★			U_2	★	★	☆	
	U_3	☆	☆	★			U_3	☆	☆	★	
4	U_1	★	★	★	×	10	U_1	★	★	★	√
	U_2	★	★	★			U_2	★	★	☆	
	U_3	☆	☆	☆			U_3	☆	★	★	
5	U_1	★	★	★	×	11	U_1	★	★	★	√
	U_2	★	★	☆			U_2	★	★	★	
	U_3	☆	☆	☆			U_3	☆	★	☆	
6	U_1	★	★	★	×	12	U_1	★	★	★	√
	U_2	★	☆	☆			U_2	★	★	★	
	U_3	★	☆	☆			U_3	★	★	★	

Note: ★ = U_i can hear A_j; ☆ = U_i cannot hear A_j; √ = the users can be localized; × = the users cannot be localized.

with circles A_2 and A_3 are $\{U_2, U_2'\}$ and $\{U_3, U_3'\}$, respectively. U_2 and U_3 can be positioned for $U_2U_3 \neq U_2'U_3'$, as shown in Figure 7.5. Therefore, the users can be localized in this situation.

3. If both U_1 and U_2 are not on line A_1A_2, assume that U_1 can be positioned; then, U_2 and U_3 can also be positioned. While U_1 rotates on circle A_1, with center A_1 and radius d_1, it is hard to find another group of U_2 and U_3 that can satisfy the conditions. Therefore, the users can be localized in this situation.

Theorem 7.7: If one of the users can receive the signal of two anchors, the second user can receive the signal of one of the two anchors and the third user can receive the signal of the third anchor (case 8 in Table 7.2), and then the users can be localized.

Proof: Assume that U_1 can hear A_1 and A_2, U_2 can hear A_2, and U_3 can hear A_3. Then,

1. If $d_{A1A2} = d_1 + d_2$, that is, U_1 is on line A_1A_2, then it can be localized. U_2 and U_3 can hear $\{U_1, A_1\}$ and $\{U_1, A_2\}$, respectively. Start from U_1 as the center; then the circles with radiuses d_{12} and d_{13}, respectively, intersect with circles A_1 and A_2 at points $\{U_2, U_2'\}$ and $\{U_3, U_3'\}$, respectively. U_2 and U_3 can be localized for $U_2U_3 \neq U_2'U_3'$, as shown in Figure 7.5. Therefore, the users can be positioned in this situation.

2. If $d_{A1A2} < d_1 + d_2$, that is, U_1 is not on line A_1A_2, then two possible localizations of U_1 that are symmetric with respect to line A_1A_2 can be obtained. For certain U_1, U_2, and U_3 can be localized because they can hear $\{U_1, A_1\}$ and $\{U_1, A_2\}$, respectively. Therefore, the users can be positioned in this situation.

Theorem 7.8: If two of the users can receive the signals of two anchors and the last user can receive the signal of the third anchor (case 9 in Table 7.2), then the users can be localized.

Proof: Assume that U_1 and U_2 can hear A_1 and A_2 and that U_3 can hear A_3. From the analysis in Theorem 7.8, we know that the users can also be localized in this situation. This is because the only difference between the situations in case 9 and case 8 is that U_2 can hear more anchors.

Theorem 7.9: If one of the users can receive the signals of all three anchors, and the other two users can receive the signals of different anchors (case 10 in Table 7.2), then the users can be localized.

Proof: Assume that U_1 can hear all three anchors and that U_2 and U_3 can hear one or two different anchors. U_1 can be localized because it can hear three noncollinear anchors. For U_2 and U_3,

1. If U_2 and U_3 can each hear one and only one anchor, we suppose that U_2 can hear A_1 and that U_3 can hear A_2. They can be positioned according to the analysis in Theorem 7.2. Therefore, the users can be localized in this situation.

2. If one of them can hear one and only one anchor and the other can hear two anchors, we suppose that U_2 can hear A_1 and that U_3 can hear A_2 and A_3 here. U_3 can be localized as it can hear U_1, A_2, and A_3. U_2 can also be positioned because it can hear U_1, A_1, and U_3. Thus, the users can be localized in this situation.

3. If all of them can hear two anchors, suppose that U_2 can hear A_1 and A_2 and that U_3 can hear A_2 and A_3. U_2 can be positioned because it can hear U_1, A_1, and A_2. U_3 can also be localized because it can hear U_1, A_2, and A_3. Therefore, the users can be localized in this situation.

Theorem 7.10: If two of the users can receive the signals of all three anchors and the third user can receive the signal of one and only one anchor (case 11 in Table 7.2), then the users can be localized.

Proof: Assume that U_1 and U_2 can hear all three anchors and that U_3 can hear A_2. U_1 and U_2 can be positioned as they can hear three noncollinear anchors. U_3 can also be localized as it can hear U_1, U_2, and A_2. In total, the users can be localized in this situation.

As observed from the theorems above, the users who cannot be localized by the traditional methods, in which the users must receive the signals of more than three anchors, can be localized according to the information of the crowd. Thus, the proposed method can improve the localization performance.

7.3.3 Detailed Crowd-Aided Localization Method Based on EKF

This section presents the crowd-aided localization method by extending the Kalman filter (EKF). The innovation of our proposed algorithm is that we utilize the information from not only the fixed anchors but also the users themselves. Moreover, we extend the Kalman filter to alleviate the effects of noisy environments and wireless signal instability. First, we introduce the motion model for the users. Second, we describe the measurement model for the distances. Finally, we introduce the proposed algorithm with mobile anchors by extending the Kalman filter, which is denoted as MEKF.

7.3.3.1 Motion Model

Mobile users moving through the environment are described by their localizations and velocities in the *X-Y* plane. Thus, the state of one user at time t can be described by a state vector: $x(t) = [Lx(t), Ly(t), Vx(t), Vy(t)]$, where $Lx(t)$ and $Ly(t)$ specify the *x*- and *y*-values and *Vx(t)* and *Vy(t)* are the user's speed in the *x*- and *y*-directions, respectively. Consequently, the state vector of *n* users in our proposed method can be described as follows:

$$X(t) = \left[x_1(t), x_2(t), \ldots, x_n(t) \right]^T,$$ (7.14)

where $x_i(t)$ represents the state of user *i*. *A* denotes the transpose operation. Therefore, the motion of the users can be described by

$$X\left(\frac{t}{t-1}\right) = A * X(t-1) + W(t-1),$$ (7.15)

where $W(t-1)$ represents noise in the process, which is assumed to be a white Gaussian noise sequence with a mean of zero and the covariance matrix *Q*. *A* is the state transition matrix, which maps the forward state transition from *t*−1 to *t*. It is defined as follows:

$$A = \begin{bmatrix} a & \cdots & O \\ \vdots & \ddots & \vdots \\ O & \cdots & a \end{bmatrix},$$ (7.16)

where *O* is a fourth-order matrix, all of whose elements are zero, and *a* can be described as follows because the state vector of user *i* at time *t* can be predicted at the same as that at time *t*−1 (*T* is the sampling time interval between two successive measurement times):

$$a = \begin{bmatrix} 1 & 0 & T & 0 \\ 0 & 1 & 0 & T \\ 0 & 0 & 1 & 0 \\ 0 & 0 & 0 & 1 \end{bmatrix},$$ (7.17)

7.3.3.2 Measurement Model

The measurement equation of the users at time instant *t* can be described as

$$Z(t) = f(X(t)) + V(t),$$ (7.18)

where $V(t) \sim N(0, R)$ is a white noise sequence that represents the measurement noise and $Z(t)$ is the measurement vector at time *t*, that is, the vector of the distances between the anchors and the users or between any two users. We take the square of the distances to create the measurement vector. We then have the following:

$$Z(t) = \left[D_{11}^2(t), \ldots, D_{ij}^2(t), \ldots, D_{mn}^2(t), D_{12}^2(t), \ldots, D_{jk}^2(t), \ldots, D_{(n-1)n}^2(t) \right]^{\mathrm{T}}, \qquad (7.19)$$

where $D_{ij}^2(t)$ describes the square of the distance between anchor i and user j ($i = 1$, 2, ..., m; $j = 1, 2, \ldots, n$) and $D_{jk}^2(t)$ represents the square of the distance between user j and user k ($j, k = 1, 2, \ldots, n$; and $j \neq k$), both of which can be described as follows:

$$D_{ij}^2(t) = \left(L_{jx}(t) - A_{ix} \right)^2 + \left(L_{jy} - A_{iy} \right)^2 + V(t), \qquad (7.20)$$

$$D_{jk}^2(t) = \left(L_{jx}(t) - L_{kx}(t) \right)^2 + \left(L_{jy}(t) - L_{ky}(t) \right)^2 + V(t), \qquad (7.21)$$

Algorithm 7.4 Interoperable Localization Based on EKF

Input: The distances D_{ij} between anchor i and user j ($i = 1, 2, \ldots, m, j = 1, 2, \ldots, n$) and the distances D_{jk} between the nonlocalized user j and the nonlocalized user k ($j, k = 1, 2, \ldots, n$, and $j \neq k$);
Output: The locations of the nonlocalized users;
1: Set the state $X\_p(0)$ and the error covariance $P\_p(0)$ all initially to 0. Initialize the predicted error Q and the measurement error R;
2: **for** the times of the localization $t := 1$ to T **do**
3: Predict the state $X\_p(t/t-1)$ according to Formula (7.3) and the error covariance by $P\_p(t/t-1) = A * P\_p(t-1) * AT + Q(t-1)$ at time $t-1$;
4: Calculate the predicted matrix h_Xp according to the positions of the anchors and the predicted state;
5: Calculate the residual $Y\_e = D_{ij}^2 = h\_X\,p$ between the actual measurement and the predicted matrix;
6: Compute the Kalman Gain $K(t) = P\_p(t/t-1) * H * ((H * P\_p(t/t-1) * H^T))^{-1}$;
7: Correct the predicted state estimate $X\_p(t) = X\_p(t/t-1) + K(t) * Y\_e$ and error covariance $P\_p(t) = [eye(length(X\_p))] * P\_p(t/t-1)$;
8: **end for**

where A_{ix} and A_{iy} are the x- and y-coordinates of anchor i, respectively ($i = 1, 2, \ldots, m$). Here, $L_{jx}(t)$ and $L_{jy}(t)$ represent the x- and y-coordinates of user j at time t, respectively ($j = 1, 2, \ldots, n$).

7.3.3.3 Interoperable Localization Algorithm

This section describes the proposed algorithm (Algorithm 7.3). The extended Kalman filter operates recursively on streams of noise to produce a statistically optimal estimate of the locations of users. In more detail, before the recursive localization process begins, the state $X\_p(0)$, the error covariance $P\_p(0)$, the predicted error Q, and the measurement error R are initialized. For each recursive process, the prior state estimate of the users, $X\_p(t/t-1)$, is measured at time $t-1$ through the nonlinear function (Equation 7.16) to measure the *a priori* state estimate at time *t*. Then, the *a priori* estimate error covariance $P\_p(t/t-1)$ at time *t* is also calculated as follows:

$$P\_p(t / t - 1) = A * P\_p(t - 1) * A^T + Q(t - 1), \tag{7.22}$$

The measurement innovation, or the residual Y_e, is calculated, which reflects the discrepancy between the actual measurement and the predicted matrix, and the Kalman gain $K(t)$ will also be computed as follows:

$$K(t) = P_p(t / t - 1) * H * (H * P_p(t / t - 1) * H^T)^{-1} \tag{7.23}$$

where $H(t)$ is the partial derivative matrix of function h:

$$H(t) = \frac{\partial h}{\partial X} | X(t / (t - 1)), \tag{7.24}$$

Finally, we use Equation 7.25 to update the *a posteriori* state estimate $X\_p(t)$ at time *t* with the calculated Kalman gain $K(t)$ in Equation 7.23.

$$X\_p(t) = X\_p(t / t - 1) + K(t) * Y\_e, \tag{7.25}$$

When the calculation is complete at time *t*, we update the *a posteriori* estimate error covariance $P\_p(t)$ to estimate the next positions of the mobile users as follows:

$$P\_p(t) = \text{eye}(\text{length}(X\_p)) * P\_p(t / t - 1), \tag{7.26}$$

For the next time instant, the preceding processes will be conducted again to calculate the new positions. The time complexity of EKF is $O(n)$, where n is the number of nonlocalized users. For our proposed MEKF algorithm, the time complexity of each localizing process is $O(1)$. It turns out that our proposed method can reduce the time complexity of the localization process.

7.3.4 Experiments

To demonstrate the performance of our proposed algorithm, extensive simulation experiments are conducted. The experiments and results are introduced in this section. Our experimental environments are as follows: In the simulation scenario, which is 18×24 m^2, eight tags were deployed. Four of them are used as the fixed anchors, which are placed at the four corners of the region. The other four tags act as the nonlocalized users: the first one is located at (10 m, 12 m), the second one walks along the edges of the rectangular experimental area, and the other two walk along both the edges and the diagonals. Figure 7.12 shows a snapshot of the

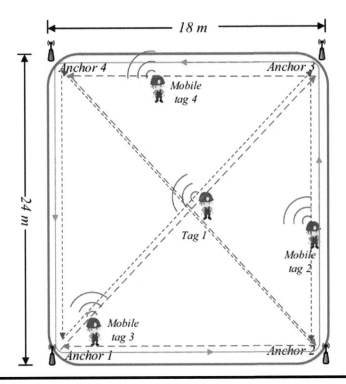

Figure 7.12   The experimental scenario.

experimental area. All these tags communicate with each other every second to calculate the distances between them, which will be used for further calculation with the proposed algorithm. Some of the main parameters for this system are listed in Table 7.3. The initial state of the nonlocalized users is $X\_p(0) = [0, 0, 0, 0, 0, 0, \ldots, 0]^T$, which means that all the users are located at the position (0, 0) and that their initial speed is 0 in both the x- and y-directions. In addition, Q and R are set to be $10^{-2} \times I$ and $10^3 \times I$, respectively, which are determined experimentally, and I denotes an identity matrix.

Figures 7.13 through 7.16 show the localization results of the four users. The figures show the moving traces of the mobile users. In Figure 7.13, the average

Table 7.3   Experimental Parameters

Parameters Values
Area size (m²): 18×24
Anchor number: 4
User number: 4
Communication range: 20
Update frequency, T (s): 1

localization errors achieved by Tri, EKF, and MEKF are 1.1218 m, 0.6632 m, and 0.5259 m, respectively, while in Figure 7.14 they are 1.4868 m, 0.9230 m, and 0.7534 m, respectively. In Figure 7.15, they are 1.3454 m, 0.8825 m, and 0.6975 m, respectively. In Figure 7.16, they are 1.3409 m, 0.9778 m, and 0.7970 m, respectively. Consequently, compared with the Tri method, MEKF can improve the localization accuracy by approximately 53.1%, 49.3%, 48.2%, and 40.6%, respectively, while compared with EKF, it improves the accuracy by 20.7%, 18.4%, 21.0%, and 18.5%, respectively. The basic reason is that the introduced mobile users can act as mobile anchors, which increases the number of available anchors.

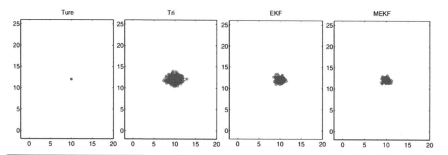

Figure 7.13   The localization results for user 1.

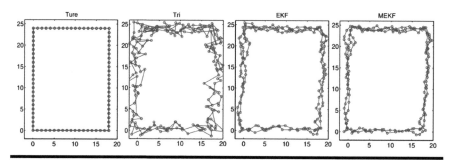

Figure 7.14   The localization results for user 2.

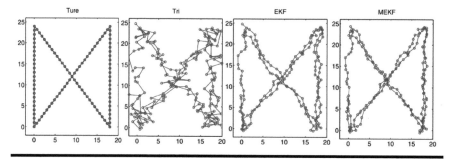

Figure 7.15   The localization results for user 3

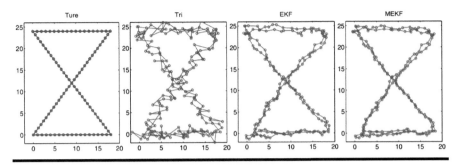

Figure 7.16   The localization results for user 4.

For more detail, Figure 7.17a presents the average localization errors of these four users as achieved by the three algorithms. From these average localization errors, we can verify that the localization accuracies of all the users are improved. This is because additional reference information can be obtained according to our proposed method, and can alleviate the influence of noise and signal instability. Therefore, the proposed method can yield more accurate localization results.

Figure 7.17b and c describes the max localization errors and their variance, respectively. As Figure 7.17b shows, the max localization errors achieved by MEKF are lower than those of Tri by 30.6%, 39.3%, 41.3%, and 32.8% for users 1–4, respectively, and lower than those of EKF by 71.8%, 38.0%, 12.8%, and 12.8%, respectively. Moreover, from Figure 7.17c, the variance of the localization errors when using MEKF is below that of Tri by 71.4%, 64.9%, 58.1%, and 46.0%, respectively, and below that of EKF by 65.3%, 40.9%, 31.4%, and 30.3%, respectively. All these results show that our proposed method with mobile anchors can achieve higher localization accuracy and more stable performance than Tri and EKF.

7.4 Direction for Future Research Works

7.4.1 Sensing Quality Evaluation

This is a problem that involves many factors. First, the distribution of crowds can affect the quality of sensing. Due to the aggregative nature of crowds, the data sensed by them only represent the information of the area with most of the people. However, such information cannot reflect the whole features of a particular location. For example, the temperature sensing task performed by crowds can only reflect the temperature information where the majority of people are. Hence, new mechanisms are proposed to minimize the negative influences. Second, the sensing ability of each device can affect the quality of sensing due to the quality of the sensors, network delay, or storage capacity. It is important to distinguish good data from poor-quality data.

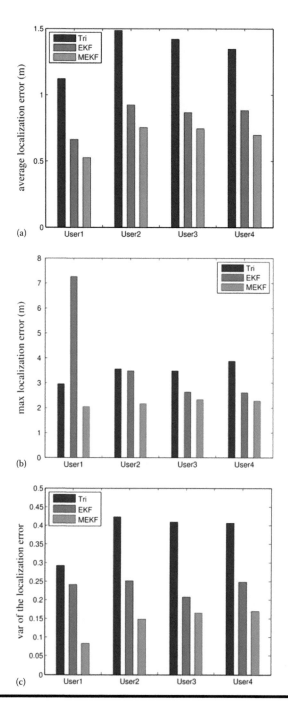

Figure 7.17 (a) The average localization error for the users. (b) The maximum localization error for the users. (c) The variance of the localization error.

7.4.2 Wireless Charging

Battery capacity is always the bottleneck limiting the lifetime of WSNs and mobile devices. The development of wireless charging techniques and the emergence of ambient backscatter have shed light on these problems. These two techniques can continually provide energy for sensor nodes or mobile crowds. However, many issues still remained unsolved in this area of combination between the mobile crowd and the charging techniques mentioned above.

7.4.3 Security Issues

With the combination of WSNs and crowds, many security issues that are nonexistent in traditional WSNs will emerge and some traditional problems will change. For example, the worm attack in MSs may cause larger damage due to the rapid spread of worms using the mobility of MSs. On the other hand, the crowd can help to alleviate or handle some security issues with their stronger computational power.

7.5 Conclusion

The development of cloud computing brings new technologies to traditional WSNs, such as high-speed computing power, big data storage, and remote service, which makes some applications possible in our lifetime. It is an inexorable trend to integrate WSNs with cloud computing. With the cloud computing paradigm adopted in WSNs, the performance of WSNs can be improved, such as energy consumption, computing latency, and service quality. However, due to the weak communication ability of WSNs, how to upload the mass sensed data to the cloud within the limited time becomes a bottleneck in the sensor–cloud integration system. Most traditional methods of mobile data collection are delay tolerant, which is not appropriate for data collection in sensor–cloud integration. In this chapter, we proposed two novel application of WSNs with the aid of crowds. The first one is TASA, which is a polynomial time algorithm to address the transmission delay problem in the data collection from WSNs to the cloud with multiple mobile crowds. The performance of the proposed method is also validated through simulations. Simulation results demonstrate that the proposed algorithm can adjust the delivery delay, reduce energy consumption significantly, and improve the system's sustainability, which contributes to the integration of WSNs.

The second one is a crowd-aided localization method for localizing mobile group users. We proposed localizing mobile users with crowds, in which the information exchanged among these users can be exploited to assist with localization. That is, localize mobile users use their interrelated information. The reason why the proposed crowd-aided localization method can improve localization probability and accuracy was given by theoretical analyses. We then extended the Kalman filter to alleviate the effects of environmental noise and signal instability. Finally, we validated the performance of the proposed localization scheme by extensive simulation experiments. The results showed that our approach significantly improved the localization accuracy and scaled quite well with the data loss rate, communication range of the users, and distance measurement error.

In the future, with the technical development of mobile devices, the applications with crowds will increasingly grow. For their mobility, strong ability, and interoperability, crowds can help do more complex work than traditional WSNs. For example, besides data collection and localization, crowds may also be used to address security problems, which always exist in both of the WSN and cloud computing fields.

References

1. T. Wang, J. Zhou, X. Chen, G. Wang, A. Liu, and Y. Liu, "A Three-Layer Privacy Preserving Cloud Storage Scheme Based on Computational Intelligence in Fog Computing," *IEEE Transactions on Emerging Topics in Computational Intelligence*, doi. org/10.1109/TETCI.2017.2764109.
2. G. Zhang, T. Wang, M. Z. A. Bhuiyan, and G. Wang, "A Fog-Based Hierarchical Trust Mechanism for Sensor-Cloud Underlying Structure," in *Proceedings of the 15th IEEE International Symposium on Parallel and Distributed Processing with Applications (IEEE ISPA 2017)*, Guangzhou, China, December 12–15, 2017.
3. T. Wang, J. Zeng, Y. Lai, Y. Cai, H. Tian, Y. Chen, and B. Wang, "Data Collection from WSNs to the Cloud Based on Mobile Fog Elements," *Future Generation Computer Systems*, doi.org/10.1016/j.future.2017.07.031.
4. M. Z. A. Bhuiyan, G. Wang, J. Wu, J. Cao, X. Liu, and T. Wang, "Dependable Structural Health Monitoring Using Wireless Sensor Networks," *IEEE Transactions on Dependable and Secure Computing*, vol. 14, no. 4, pp. 363–376, 2017.
5. T. Wang, Z. Peng, S. Wen, Y. Lai, W. Jia, Y. Cai, H. Tian, and Y. Chen, "Reliable Wireless Connections for Fast-Moving Rail Users Based on a Chained Fog Structure," *Information Sciences*, vol. 379, pp. 160–176, 2017.
6. J. Shen, H. W. Tan, J. Wang, J. W. Wang, and S. Y. Lee, "A Novel Routing Protocol Providing Good Transmission Reliability in Underwater Sensor Networks," *Journal of Internet Technology*, vol. 16, no. 1, pp. 171–178, 2015.
7. D. V. Jose and G. Sadashivappa, "A Novel Scheme for Energy Enhancement in Wireless Sensor Networks," in *Computation of Power, Energy Information and Communication (ICCPEIC), 2015 International Conference on*, IEEE, New York, 2015, pp. 0104–0109.
8. M. Z. A. Bhuiyan, J. Wu, G. M. Weiss, T. Hayajneh, T. Wang, and G. J. Wang, "Event Detection through Differential Pattern Mining in Cyber-Physical Systems," *IEEE Transactions on Big Data*, https://doi.org/10.1109/TBDATA.2017.2731838.
9. M. Z. A. Bhuiyan, J. Wu, G. Wang, and T. Wang, and M. Hassan, "e-Sampling: Event-Sensitive Autonomous Adaptive Sensing and Low-Cost Monitoring in Networked Sensing Systems," *ACM Transactions on Autonomous and Adaptive Systems*, vol. 12, no. 1, pp. 1–29, 2017.

10. M. Z. A. Bhuiyan, J. Wu, G. Wang, Z. Chen, J. Chen, and T. Wang, "Quality Guaranteed and Event-Sensitive Data Collection and Monitoring in Wireless Vibration Sensor Networks," *IEEE Transactions on Industrial Informatics*, vol. 13, no. 2, pp. 572–583, 2017.

11. R. Zhang, J. Pan, D. Xie, and F. Wang, "NDCMC: A Hybrid Data Collection Approach for Large-Scale WSNs Using Mobile Element and Hierarchical Clustering," *IEEE Internet of Things Journal*, vol. 3, no. 4, pp. 533–543, 2016.

12. F. Restuccia, G. Anastasi, M. Conti, and S. K. Das, "Analysis and Optimization of a Protocol for Mobile Element Discovery in Sensor Networks," *IEEE Transactions on Mobile Computing*, vol. 13, no. 9, pp. 1942–1954, 2014.

13. J. S. Prashanth and S. V. Nandury, "Cluster-Based Rendezvous Points Selection for Reducing Tour Length of Mobile Element in WSN," in *Advance Computing Conference (IACC), 2015 IEEE International*. New York: IEEE, 2015, pp. 1230–1235.

14. J. J. Chen, C. F. Xing, Z. F. Cheng, and X. B. Wang, "Node Information Collection Routing Algorithm Using Mobile Agent in WSN Initialization Stage," *Applied Mechanics and Materials*, vol. 462–463, pp. 64–67, 2013.

15. T. Suetsugu, S. Matsunaga, T. Torikai, and H. Furukawa, "Effective Data Collection Scheme by Mobile Agent Over Wireless Sensor Network," *IEICE Transactions on Communications*, vol. 2015, no. 3, pp. 1–6, 2015.

16. M. Dong, K. Ota, L. T. Yang, S. Chang, H. Zhu, and Z. Zhou, "Mobile Agent-Based Energy-Aware and User-Centric Data Collection in Wireless Sensor Networks," *Computer Networks*, vol. 74, no. PB, pp. 58–70, 2014.

17. K. Trivedi and A. K. Srivastava, "An Energy Efficient Framework for Detection and Monitoring of Forest Fire Using Mobile Agent in Wireless Sensor Networks," in *IEEE International Conference on Computational Intelligence and Computing Research*. New York: IEEE, 2015, pp. 1–4.

18. M. Arquam, C. Gupta, and M. Amjad, "Delay Constrained Routing Algorithm for WSN with Mobile Sink," in *Computational Science and Engineering (CSE), 2014 IEEE 17th International Conference on*. New York: IEEE, 2014, pp. 1449–1454.

19. G. Hou, X. Wu, C. Huang, and Z. Xu, "A New Efficient Path Design Algorithm for Wireless Sensor Networks with a Mobile Sink," in the *27th Chinese Control and Decision Conference (2015 CCDC)*. New York: IEEE, 2015, pp. 5972–5977.

20. T. Wang, W. Jia, B. Zhong, H. Tian, and G. Zhang, "Blue Cat: An Infrastructure-Free System for Relative Mobile Localization." *Adhoc & Sensor Wireless Networks*, vol. 29, no. 1–4, p. 133, 2015.

21. J. Wang, Y. Zhang, Z. Cheng, and X. Zhu, "EMIP: Energy-Efficient Itinerary Planning for Multiple Mobile Agents in Wireless Sensor Network," *Telecommunication Systems*, vol. 62, no. 1, pp. 93–100, 2016.

22. J. Shi, X. Wei, and W. Zhu, "An Efficient Algorithm for Energy Management in Wireless Sensor Networks via Employing Multiple Mobile Sinks," *International Journal of Distributed Sensor Networks*, vol. 12, no. 1, pp. 1–9, 2016.

23. C. Zhu, V. C. Leung, L. T. Yang, and L. Shu, "Collaborative Location-Based Sleep Scheduling for Wireless Sensor Networks Integrated with Mobile Cloud Computing," *IEEE Transactions on Computers*, vol. 64, no. 7, pp. 1844–1856, 2015.

24. Y. Li, T. Wang, G. Wang, J. Liang, and H. Chen, "Efficient Data Collection in Sensor-Cloud System with Multiple Mobile Sinks," in *Advances in Services Computing: 10th Asia-Pacific Services Computing Conference, APSCC 2016*, Zhangjiajie, China, November 16–18, 2016, *Proceedings*. Berlin: Springer, 2016, pp. 130–143.

25. T. Wang, Z. Peng, W. Xu, J. Liang, G. Wang, H. Tian, Y. Cai, and Y. Chen, "Cascading Target Tracking Control in Wireless Camera Sensor and Actuator Networks," *Asian Journal of Control*, doi.org/10.1002/asjc.1525.

26. T. Wang, J. Zeng, M. Z. A. Bhuiyan, H. Tian, Y. Cai, Y. Chen, and B. Zhong. "Trajectory Privacy Preservation Based on a Fog Structure in Cloud Location Services," *IEEE Access*, vol. 5, no. 1, pp. 7692–7701, 2017.

27. T. Wang, Z. Peng, J. Liang, S. Wen, M. Z. A. Bhuiyan, Y. Cai, and J. Cao, "Following Targets for Mobile Tracking in Wireless Sensor Networks," *ACM Transactions on Sensor Networks*, vol. 12, no. 4, pp. 31.1–31.24, 2016.

28. A. A. Khudhair, S. Q. Jabbar, M. Q. Sulttan, and W. Deshengt, "Wireless Indoor Localization Systems and Techniques: Survey and Comparative Study," *Indonesian Journal of Electrical Engineering and Computer Science*, vol. 3, no. 2, pp. 392–409, 2016.

29. H. Chenji and R. Stoleru, "Toward Accurate Mobile Sensor Network Localization in Noisy Environments," *IEEE Transactions on Mobile Computing*, vol. 12, no. 6, pp. 1094–1106, 2013.

30. S. Van de Velde, G. T. de Abreu, and H. Steendam, "Improved Censoring and NLOS Avoidance for Wireless Localization in Dense Networks," *IEEE Journal on Selected Areas in Communications*, vol. 33, no. 11, pp. 2302–2312, 2015.

31. S. Salari, S. Shahbazpanahi, and K. Ozdemir, "Mobility-Aided Wireless Sensor Network Localization via Semidefinite Programming," *IEEE Transactions on Wireless Communications*, vol. 12, no. 12, pp. 5966–5978, 2013.

32. P. Biswas, T. C. Liang, K. C. Toh, Y. Ye, and T. C. Wang, "Semidefinite Programming Approaches for Sensor Network Localization with Noisy Distance Measurements," *IEEE Transactions on Automation Science and Engineering*, vol. 3, no. 4, pp. 360–371, 2006.

33. Y. Liu, Y. Han, Z. Yang, and H. Wu, "Efficient Data Query in Intermittently-Connected Mobile Ad Hoc Social Networks," *IEEE Transactions on Parallel and Distributed Systems*, vol. 26, no. 5, pp. 1301–1312, 2015.

34. Y. Liu, H. Wu, Y. Xia, Y. Wang, F. Li, and P. Yang, "Optimal Online Data Dissemination for Resource Constrained Mobile Opportunistic Networks," *IEEE Transactions on Vehicular Technology*, vol. 66, no. 6, pp. 5301–5315, 2017.

35. Y. Zhang, X. Sun, and B. Wang, "Efficient Algorithm for K-Barrier Coverage Based on Integer Linear Programming," *China Communication*, vol. 13, no. 7, pp. 16–23, 2016.

36. C. Y. Chang, T. L. Wang, and C. Y. Tung, "A Mobile Anchor Assisted Localization Mechanism for Wireless Sensor Networks," in *Wireless Communications and Networking Conference (WCNC), 2014 IEEE*. New York: IEEE, 2014, pp. 2793–2798.

37. H. Bao, B. Zhang, C. Li, and Z. Yao, "Mobile Anchor Assisted Particle Swarm Optimization (PSO) Based Localization Algorithms for Wireless Sensor Networks," *Wireless Communications and Mobile Computing*, vol. 12, no. 15, pp. 1313–1325, 2012.

38. S. Halder and A. Ghosal, "A Survey on Mobile Anchor Assisted Localization Techniques in Wireless Sensor Networks," *Wireless Networks*, vol. 22, no. 7, pp. 2317–2336, 2016.

39. M. Z. Win, A. Conti, S. Mazuelas, Y. Shen, W. M. Gifford, D. Dardari, and M. Chiani, "Network Localization and Navigation via Cooperation," *IEEE Communications Magazine*, vol. 49, no. 5, pp. 56–62, 2011.

40. A. Conti, M. Guerra, D. Dardari, N. Decarli, and M. Z. Win, "Network Experimentation for Cooperative Localization," *IEEE Journal on Selected Areas in Communications*, vol. 30, no. 2, pp. 467–475, 2012.

41. H. Wymeersch, J. Lien, and M. Z. Win, "Cooperative Localization in Wireless Networks," *Proceedings of the IEEE*, vol. 97, no. 2, pp. 427–450, 2009.

42. N. Patwari, J. N. Ash, S. Kyperountas, A. O. Hero, R. L. Moses, and N. S. Correal, "Locating the Nodes: Cooperative Localization in Wireless Sensor Networks," *IEEE Signal Processing Magazine*, vol. 22, no. 4, pp. 54–69, 2005.

43. R. W. Ouyang, A. K. S. Wong, and C. T. Lea, "Received Signal Strength-Based Wireless Localization via Semidefinite Programming: Noncooperative and Cooperative Schemes," *IEEE Transactions on Vehicular Technology*, vol. 59, no. 3, pp. 1307–1318, 2010.

Chapter 8

A Conditional Privacy Framework for Crowd-Assisted Community Policing

Avinash Srinivasan and Mario Gerla

Contents

8.1 Introduction

Fundamental advances in the computing and communications industry have catalyzed the transformation of crowd computing from a mere plausibility to a hard reality. It has fueled numerous business opportunities in the service industry as well as innovation across industry verticals that were previously beyond the reach of contemporary computing resources. It is now time to embrace the dawn of such an evolution to develop innovative solutions to address the ever-growing and seemingly insurmountable challenge faced by the law enforcement (LE) community.

Crowd computing—crowd sourcing (see definition 1) and crowd sensing (see definition 2)—at its core, has the potential to solve extremely large and complex problems by harnessing limitless resources and the vast knowledge of citizens with mobile sensing and computing devices. It is now time to explore the feasibility of community and LE collaboratively working to effectively and quickly solve crime. Conventional wisdom has society believing that LE is solely responsible for solving crime. While at the heart of the judicial system this is true, LE agencies have very limited resources, especially field personnel. Consequently, it is neither possible nor practical for LE personnel to monitoring all places at all times.

On the contrary, it is almost always true that few citizens are bound to be in any specific location of interest to LE after an incident. There is always a need for service-minded citizens to step up and volunteer information to help LE in promoting peace within the community. Therefore, the investigation process itself can benefit in numerous ways from participatory sensing and computing. Especially, the recent advancement in technology enables citizens to sense significant amounts of data that go unused. However, citizens' fear of retaliation primarily prevents them from participating in community policing.

The U.S. Department of Justice (DoJ) community policing initiative is a multi-objective initiative to promote a healthy relationship between LE agencies and communities. This is particularly important to citizens residing in or transiting through areas with high rates of crimes, such as drive-by shootings, gang-related violence, vandalism, random acts of violence, drug pedaling, and hate crimes, among others.

Reducing the fear of crime in the community and its citizens has always been a key focus of judiciary systems worldwide. Such fear can negatively impact the quality of life of people far beyond areas directly affected by any specific event. Unchecked fear of such crimes over extended periods can quickly manifest in numerous other secondary crimes. One specific secondary crime that is a major concern is hate crimes perpetrated against a broad race or ethnicity.

8.1.1 Current Challenges to Effectively Utilizing Crowd-Sensed Data

The Internet edge devices, such as smartphones, tablets, GPSs, and in-vehicle sensors, continue to be leveraged to feed sensed data to the Internet on a societal scale. From such large amounts of edge data fed from mobiles that are constantly moving and sensing, extremely useful opportunistic data can be harvested. However, most of the data sensed by mobiles, such as vehicles and smartphones, primarily stays on the device, with the exception of occasional events that trigger YouTube or Facebook uploads or tweets or blogs. This data can be a potential gold mine for various applications, especially to LE in solving extremely time-sensitive crimes such as a kidnapping or hit-and-run.

However, the lack of a well-designed framework with robust privacy enforcement mechanisms hinders the effective utilization of unfathomable quantities of crowd-sensed multimodal data. To make matters worse, there is a deep-rooted fear among citizens to provide any information or assistance to LE. There is a multifaceted perceived fear of reprisal, getting dragged into convoluted legal matters, retaliation, or other unwanted consequences, such as the targeted enforcement of speeding and, insurance premium hikes. Even when users are willing to participate in applications requiring community sensing that can improve their safety and security, they have very strong reservations about their identity and privacy. There is a lack of a robust framework that can identify, correlate, and extract relevant information in a privacy-preserving manner. Hence, there is an urgent need to design a well-structured approach to systematically harness meaningful data from the crowd-sensed data streams (Table 8.1).

In light of the above-discussed facts, in this chapter we propose a crowd-assisted application platform that can provide granular privacy control with the objective of harnessing the untapped potential of community-oriented data sensing. We will examine the impact of a conditional privacy-preserving community policing framework that leverages crowd assistance in time-sensitive crime investigations. The proposed framework addresses issues of agency-specific privacy policies, regulatory compliance, privacy laws, and user privacy expectations with granular controls. The framework is designed with strong cryptographic primitives that are provably secure. Furthermore, LE can provision new cloud-oriented applications, such as *sensing-as-a-service*, to support information and intelligence gathering as needed.

Three broadly accepted approaches among researchers and practitioners for privacy preservation in crowd-based applications are *anonymization*, *encryption*, and *data perturbation*. Our proposed framework leverages anonymization and encryption for enforcing granular privacy controls specific to the domain requirements.

Table 8.1 Summary of Multimodal Crowd-Sensed Data

Communication Modality	Notation	Useful Data
Infrastructure to mobile	I2M	Google Maps, Waze, Amber Alerts, etc.
Infrastructure to vehicle	I2V	Advisory broadcast by RSU
Infrastructure to infrastructure	I2I	Advisory broadcast by RSU
Mobile to infrastructure	M2I	Cell phone-to-tower communication
Mobile to mobile	M2M	Find My Friends, file sharing, etc.
Mobile to vehicle	M2V	Apps and Bluetooth data
Vehicle to infrastructure	V2I	Vehicle reports traffic to RSUs
Vehicle to mobile	V2M	Mobile phone Bluetooth log
Vehicle to vehicle	V2V	Emergency breaking, lane change, etc.
On-board vehicle data	V_{data}	Infotainment and telematics data, OBU data, etc.

Note: RSU = roadside unit; OBU = on-board unit.

8.1.2 *Summary of Contributions*

In this chapter, we propose a solution approach that can provide a secure, distributed, privacy-preserving, crowd-assisted framework for crime investigations. To the best of our knowledge, the proposed framework is the first formal work to systematically address the challenges and requirements of crowd-assisted crime investigations. This is also the first work to present details of an anonymity-preserving solution architecture within the context of crowd-assisted crime intelligence garnering.

Our proposed framework provides a community policing platform with robust identity privacy against unauthorized use and disclosure. The solution framework promotes anonymous community group subscription mechanisms using strong and provably secure cryptographic primitives, such as ring signatures. The community policing platform provides users with an easy-to-use web-based user interface (UI) to submit information useful to LE in a privacy-preserving manner.

One of the most unique aspects of this framework is vesting privacy controls in the hands of the data owner. This is achieved through the use of simple and easy-to-understand high-level privacy policies, which has two key outcomes.

First, it enables data owners to be more actively involved in protecting their privacy. Second, it promotes enhanced awareness of privacy requirements and the consequences of privacy breaches. Thus, it offers distributed, anonymous, data query posting mechanisms, including the web, social networks, and public transportation for multimodal network configurations (wired, wireless, and opportunistic).

8.1.3 Chapter Organization

The remainder of this chapter is organized as follows. In Section 8.2, we provide a discussion on the relevant background and related contemporary works. This section also presents the broad scope of the problem and the motivation of this chapter. In Section 8.3, we present the preliminaries and requirements for the proposed community policing framework. In Section 8.4, we describe the proposed crowd-based community policing framework, followed by a detailed use case intuitively describing the usefulness of the proposed solution framework in Section 8.5. Finally, we conclude this chapter in Section 8.6 by highlighting some of the key directions for future research in this domain.

8.2 Background and Related Work

Crowd-assisted computing can be categorized into two broad categories: *crowd sourcing* and *crowd sensing*.

> **Definition 1.** *Crowd sourcing as a computing paradigm engages citizens to collectively solve a problem or challenge at hand, primarily via online communities.*
> **Definition 2.** *Crowd sensing as a new business model allows a large number of mobile computing devices to collect data with the same end goal.*

Crowd sourcing approaches have been studied in a variety of related projects to harness large bodies of human resources, to generate data for use in several systems, and to complete tasks that are costly or time-consuming with traditional methods [1]. In their work [1], Yan et al. took advantage of the emergence of smartphones and implemented an iOS mobile application, mCrowd, to distribute image-related tasks. mCrowd uses the Amazon Mechanical Turk (MTurk) and ChaCha crowd sourcing platforms to perform tasks such as querying images and capturing geo-tagged images, among other image-related tasks.

Ambati et al. present Active Crowd Translation (ACT) in [2], which is a system that leverages crowd sourcing via Amazon MTurk to aid in machine translation for language pairs. CrowdDB [3] was developed to leverage human input to respond to database queries that cannot be otherwise handled by the database management system.

Dehghantanha and Franke discuss privacy-respecting digital forensics as an emerging cross-disciplinary research area in [4]. Their work attempts to describe the details of "privacy-respecting digital investigation" as a cross-disciplinary field of research. Finally, they look at potential privacy issues during digital investigation in the light of EU, U.S., and Asia-Pacific Economic Cooperation (APEC) privacy regulations.

Current state-of-the-art tools available to LE include the Law Enforcement Alerting Portal (LEAP) and America's Missing: Broadcast Emergency Response (AMBER) alerts. LEAP is the web portal that is used by LE to enter the missing person's information. This information is then used to create AMBER alerts for children 17 and younger and notify the public through multiple broadcast avenues, such as cellular service providers, highway signboards, advertising billboards, and radio.

One other work that talks about leveraging the masses for solving old and cold cases, CrowdSolve, is presented in [5]. The app seeks to allow a mass audience to review the documents and data surrounding criminal cases in the hopes that someone may find information otherwise missed by the investigators. Nonetheless, there are no details about how the app works, especially on how user privacy is protected. Furthermore, there is no discussion on how case information will be sanitized in compliance with a wide array of laws and regulations before it is shared with the masses. These may be a few reasons that the CrowdSolve app was considered unsuccessful.

Our proposed solution framework can be integrated into the LEAP/AMBER system to enhance its capabilities further. This framework facilitates both real-time and after-the-fact crowd assistance in a secure and privacy-preserving manner. Through identity-preserving tools, LE can seek information from the masses, especially since users can provide information controlling their privacy granularity. For example, if the vehicle identified in the ABER alert is spotted by a citizen, then the information provided to LE can include the fine-grain GPS location of the reporting vehicle or a high-grain location within an x-mile radius.

Bender et al. [6] proposed a ring signature scheme based on stronger definitions of security. Their proposed scheme included a ring signature construction without oracles. Several more generic ring signature schemes have been proposed over the years based on various assumptions [7–9]. Some of the efficient ring signature schemes proposed over the years include [10–12]. In [13], Fujisaki and Suzuki introduce a ring signature scheme specifically designed to make unauthorized disclosures anonymously, without the risk of identity escrow. This lack of need for identity escrow allows users of the ring signature to hide themselves under the cover a group. Consequently, this provides an inherently higher level of distrust compared with schemes relying on a group manager [14].

In all the proposed ring signature schemes with the exception of that of Dodis et al. [8], the size of the ring is assumed to be linear in the number of the members. This has been primarily because a list of members' public keys is the simplest and most natural way to describe a ring. However, Dodis et al. [8] describe their ring signature scheme with constant-size signatures. They argue that rings can be created with simple and short descriptions that are intuitive, such as "members of White House staff." Furthermore, they argue that rings can be reused. Therefore, in such scenarios, the rings need not be created fresh for every single signature, and instead could be assigned a unique description or identifier. Nonetheless, the constant size of ring signatures is achieved by relying on random oracles [15].

Everyday crimes in human society include incidents such as hit-and-run traffic accidents, child abductions, and drive-by shootings. A summary of crime data is presented in Table 8.2. Such crimes can go cold if no timely information is

Table 8.2   Statistics on Hit-and-Run, Missing Persons, and Wanted Felons

Crime Category	Current State
Hit-and-run	In 2013, one-fifth of all pedestrians killed in traffic crashes were involved in hit-and-run collisions [18]. Nearly 1500 people die annually in hit-and-run crashes; about 11% of all police-reported crashes involve at least one driver who flees the scene. An average of 96 people died each day in motor vehicle crashes in 2015, i.e., one fatality every 15 minutes. On average, a pedestrian was killed every 2 hours and injured every 7 minutes in traffic crashes [18]. Drivers involved in hit-and-runs are more likely to be young (under the age of 25), male, and intoxicated, and drive a stolen vehicle, often without a valid license [19,20].
Kidnapping and missing persons	The FBI's National Crime Information Center's (NCIC) annual numbers of missing person cases handled are as follows: 2013—627; 911 (T), 462; 567 (J) [21] 2014—635; 155 (T), 466; 946 (J) [22] 2015—637; 304 (T), 442; 442 (J) [23] 2016—647; 435 (T), 447; 444 (J) [24] As of December 31, 2016, the NCIC contained 88,040 active missing person records. Juveniles aged <18 years account for 38.3% of these records, and 48.6% hen juveniles are defined as <21 years of age [24].
Wanted felons	According to a 2013 U.S. Bureau of Justice Statistics, a combined total of ≈2000 state and federal inmates escaped or went off without approval.

Note: T = total; J = juvenile.

generated that can lead investigators in the right direction. The initial time window, which can vary from several minutes to the first 48 hours, depending on the crime, is extremely crucial. However, LE has limited personnel and resources, which leaves a wide window of opportunity in time and space (geographical) for the perpetrator to slip beyond the reach of the justice system.

In such scenarios, if the power of community policing and crowd sensing is leveraged appropriately, it exponentially augments the apprehension of the perpetrators. On the flip side, in community sensing applications, privacy concerns are plentiful and limit the access LE has to these crowd-sensed data streams. Therefore, it is critical that a reliable crowd sensing application platform is developed that citizens trust to satisfy their privacy requirements.

8.3 Framework Requirements and Preliminaries

8.3.1 Anonymous Participation

Anonymous participation will enhance privacy, and consequently, users will feel more confident and assured about their privacy. They will thus have little resistance to participate in community policing. The lack of robust privacy enforcement mechanisms can potentially lead to retaliation and even jeopardize the users' personal safety. This is at the heart of the current lack of effective cooperation and strained relations between citizens and LE.

> **Definition 3.** *Anonymity is the state of being not identifiable within a set of subjects [16].*
>
> **Definition 4.** *An anonymity set is a set of subjects with potentially the same attributes, any of which might cause an action [16].*

However, a solution that provides total anonymity would imply that there are no consequences for malicious or reckless behavior. Such a solution framework may result in users utilizing the platform for personal vendettas by providing falsified information. Such user behavior can have varied impacts on both citizens and LE, including wasted LE resources and efforts, false arrests of innocent citizens, and intentionally misleading LE with decoy intel.

Effective measures must be in place to prevent such misuse of the community policing platform. Otherwise, it will continue to deter efforts to promote safer communities and improved relationships between citizens and LE. Therefore, conditional anonymity has the potential to provide a middle ground where users enjoy anonymity as long as they are playing fair and by the rules (Figure 8.1). Suspicious behavior will result in being flagged, and repeated flags will eventually result in user(s) being deanonymized and reprimanded. The above two broad categories of anonymity are discussed in detail below.

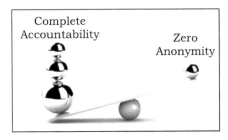
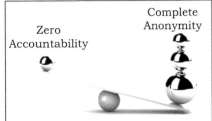
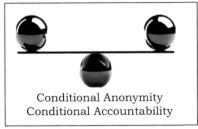

Figure 8.1 Illustration of the trade-off between anonymity and privacy, and the usefulness of conditional anonymity.

8.3.1.1 Conditional Anonymity with Group Signatures

The coexistence of anonymity and identity traceability is known as *conditional anonymity*. While conditional anonymity can provide total anonymity, it is conditional on sincere participation from users. While these systems are also susceptible to foul play from malicious actors, because the anonymity is conditional on good behavior, the malicious actions can be traced to a specific user. Therefore, malicious or reckless behavior is not free from consequences, since such users will be identified and reprimanded. Consequently, unlike unconditional anonymity systems, they cannot provide *nonrepudiation* and *accountability*.

Group signatures are a modern cryptographic primitive that allow a member of a specific group to sign messages on behalf of the group as a whole [17]. Consequently, users can sign without revealing their individual identities. This provides them with a certain degree of anonymity and privacy. More importantly, it still guarantees that the user who signs is in fact a member of the right group. Furthermore, the whole idea of a group signature rests on the assumption that a trusted group leader exists who is responsible for setting up a "user group."

The group leader also holds a master key with which he can reveal the identity of the signer of any signature generated by a group member in the past. Therefore, group signatures only provide conditional anonymity since the group leader can trace any individual signer from his group if and when needed. However, group signatures are not suitable for forming groups on an ad hoc basis.

8.3.1.2 Unconditional Anonymity with Ring Signatures

Unconditional anonymity systems provide perfect privacy. This translates into preserving member identities regardless of time, length, or frequency of communication.

> **Definition 5.** *Ring signature, a cryptographic primitive, is a type of digital signature that can be performed by any member of a group of users that each have keys. Ring signatures are primarily used for anonymous authentication of a user as a member of a group, thereby protecting the user's individual identity.*

Ring signatures (definition 5), first introduced in [25], are a complement to group signatures and were created in response to the limitations of group signatures discussed in Section 8.3.1.3. They are cryptographic protocols designed to allow any member of a group to produce a signature on behalf of the group. More importantly, the group signatures are generated without revealing the signer's individual identity. This property of ring signature schemes provides anonymous authentication. One of the security properties of a ring signature is that it should be computationally infeasible to determine which of the group members' keys was used to produce the signature. Therefore, ring signatures offer *unconditional anonymity* to group (aka ring) members that is not attainable through other generic digital signature schemes or group signature schemes.

Members of an anonymity set in ring signature schemes are afforded unconditional anonymity since they operate autonomously without the need for a ring leader or any form of central authority. A key limitation of group signature schemes overcome by ring signatures is the ability to form ad hoc rings without a complex setup procedure or the requirement for a group leader. The only *a priori* requirement for participating in ring signature schemes is that the user be part of an existing *public key infrastructure*. In summary, ring signatures differ from group signatures in two key ways: first, there is no way to revoke the anonymity of an individual signature, and second, any group of users can be used as a group without additional setup.

However, perfect privacy can only be achieved when members are afforded total anonymity. While perfect privacy is highly desirable, it survives and thrives only in systems where participants are 100% truthful all the time. Otherwise, such systems are extremely vulnerable to foul play from malicious actors, since lack of traceability implies no consequences for malicious behavior. Additionally, unconditionally anonymous systems cannot provide the following two key security requirements: *nonrepudiation* and *accountability*.

8.3.1.3 Conditional Anonymity with Deniable Ring Signatures

The above-discussed conditional anonymity with ring signatures method indeed appears to be a suitable protocol that will benefit community policing needs. However, on a closer look, such total anonymity-preserving tools create a serious asymmetry situation benefiting malicious users. Therefore, from the unique scenario of community policing, what we need is a tool that can provide unconditional anonymity to honest users, but with a trapdoor mechanism that can trace and deanonymize a malicious user. Conditional anonymity ring signatures provide the important property of *deniability*.

A conditionally anonymous ring signature (CAR) termed *deniable ring signature* was first proposed by Komano et al. [26]. This variant of the traditional ring signature does provide anonymous authentication, allowing a member to sign a message on behalf of the ring. However, the anonymity offered by deniable ring signature schemes is conditional; that is, anonymity is revocable under well-defined conditions and authority. Consequently, malicious actors could now be traced when needed. Therefore, ring signatures can provide *unconditional anonymity* or *conditional anonymity*, in addition to anonymous authentication.

The anonymous authentication has two fundamental properties that are key to its use in siphoning crime intel data for crime investigations: *ring size anonymity* and *ring member anonymity*. The analyzer, which is the LE personnel, knows that the data comes from a (typically small) group, but does not know from which member in the group. Additionally, only members of the group can participate in the group communications. The tool will provide both group-level anonymous authentication and secure group communications.

Zeng et al. have proposed a CARS scheme in [27] that has the ability to provide total anonymity with certain traceability conditions. From the citizens' viewpoint, they need assurance that the community policing platform satisfies the *unforgeability* property; that is, no one other than the members of the ring can generate a valid signature on behalf of that ring. More importantly, no authorized member of the ring can forge an authentic signature on behalf of the group that can be traced to a different group member. This property is more formally referred to as *nonframeability*.

8.4 Privacy-Preserving Community Policing Framework

Anonymity is a key requirement in information-sharing applications, especially when the information shared can have a negative impact to the provider if misused. While conventional wisdom has users believing that anonymity and accountability are at odds, cryptographic primitives have the unique ability for provisioning strong anonymity with accountability, if leveraged appropriately. An ideal anonymity-based system should encourage honest users' participation, discourage dishonest behavior, and reprimand those that misuse the anonymous system. To this aim, for the proposed community policing framework, the following requirements outlined in [28] have to be satisfied:

1. Authenticate honest users with anonymous credentials.

2. Identify malicious users under well-defined conditions leveraging conditional anonymity property.
3. Handle emergency situations utilizing revocable anonymity though group signatures [29].

8.4.1 Protocol Preliminaries

Let u be the universe of users defined in Equation 8.1.

$$u = \{u_1, u_2, \ldots, u_x\} = \{u_i \mid i = 1, 2, \ldots, x\} \tag{8.1}$$

Let each community policing initiative be denoted as an anonymity set (aka ring) r_j such that the set of all rings is denoted as shown in Equation 8.2.

$$\chi = \{r_1, r_2, \cdots, r_p\} = \sum_{j=1}^{p} r_j \tag{8.2}$$

Each ring has one master node denoted as r_j^m, responsible for executing the confirm and refute protocols, when the user has to be verified for honesty. Furthermore, $\forall u_i \in r_j$, there is a key pair ($K_{pub}^{u_i}, K_{prv}^{u_i}$), where $K_{pub}^{u_i}$ is u_i's public key and $K_{prv}^{u_i}$ is u_i's private (secret) key. The set of all public keys κ_{pub} for the ring r_j is denoted as shown in Equation 8.3.

$$\kappa_{pub}^{r_j} \leftarrow \bigcup_{j=1}^{n} K_{pub}^{u_i} \tag{8.3}$$

The original ring signature proposed by [25] consists of three algorithms, explained below.

1. *KeyGen* (κG)—generates a key pair <k_{pub}, k_{prv}>
2. *Sign* (*SG*)—generates a unique ring signature σ for an input triple of the form <m, k_{prv}, r>
3. *Verify* (*VR*)—verifies a given ring signature σ for the corresponding <m, r>; outputs 1 if signature is authentic, 0 otherwise

To achieve conditional anonymity using ring signatures, we need more efficient ring signature protocols. A key requirement of such protocols is the traceability of the message signer under well-established deanonymizing conditions. The traceability property helps introduce accountability for misbehavior. Furthermore, this property will be very beneficial if participants express consent to be summoned for witness testimony if LE is in need of a witness. Such a ring signature scheme is referred to as a deniable ring signature. In our proposed framework, let us denote the deniable ring signature scheme as follows: let SG_{drs} for a message m user $u_i \hat{I} r_x$ as σ_{drs}. In this case, KGR_{drs} generates the key pair $< k^i_{pub}, k^i_{prv} >$. VR_{drs} can authenticate σ_{drs}, and the new function *Trace* (TR_{drs}) can identify the original signer unambiguously.

8.4.2 Framework Working Details

The proposed framework should minimize the workload and installation and configuration complexity for the participants. This is vital for recruiting community residents to participate in community policing. Therefore, the proposed crowd-assisted community policing framework is envisioned to operate in a client–server environment. This will enable all the resource-intense operations to be implemented on the server side with a lot of flexibility. On the other hand, client-side elements will be very light and intuitive to use, devoid of any significant sensing, processing, or transmission of data (Figure 8.2).

Let each participating LE agency have a unique *PublicKey, PrivateKey* key pair denoted as $- K^{LE_{id}}_{pub}, K^{LE_{id}}_{prv}$, and the key pair is assigned to each LE agency by a central authority. Note that each LE agency has a unique ID, LE_{id}. Furthermore, for

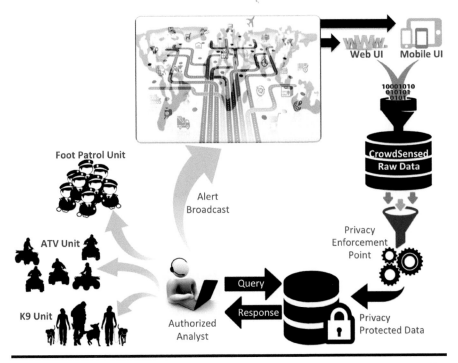

Figure 8.2 **Information flow in the proposed crowd-assisted community policing framework.**

each event for which an LE agency decides to employ crowd assistance, the agency assigns the event a unique ID, ID_{evt}.

Immediately after an event occurs, the LE first responders enter the information into a system such as LEAP and generate a unique event ID, ID_{evt}. The event is then broadcast using alerting mechanisms such as AMBER, cellular and cable television service providers, radio, roadside units, and highway billboards. The unique event ID generated by the system is included with the alert broadcast. Anonymous participation in response to this event alert is limited to users who are preregistered and have either the mobile app or valid web portal authentication credentials.

ID_{evt} is included in the report submitted by users. The user will submit information through a combination of drop-down, check-box, and text fields (both mobile app and web-based reporting). The entire report along with ID_{evt} is signed by the user with a ring signature. The signed event report is then encrypted with an appropriate LE agency public key, $K_{pub}^{LE_{id}}$. For illustration, consider the event of locating abandoned stolen vehicles or the real-time tracking of an in-progress carjacking. The above mechanism is well suited for handling crimes of this nature.

However, if the policing event is more severe, such as apprehending suspects involved in a robbery or drive-by shooting, it is hard for citizens to come forth with information unless they are ensured anonymity and protection. To this aim, LE can preregister good citizens in at-risk neighborhoods who care about the community. To do this, all they have to do is register themselves at the participating LE agency and obtain a pseudo ID. When such citizens learn about imminent threats to residents of the community or want to share information postincident, they can include their pseudo ID in the report, which can be deidentified as and when needed.

A postincident report in this scenario would be similar to Equation 8.4. The report includes event details such as event type E_{type}, event location E_{loc}, and time window t_{min} and t_{max}, as shown in Equation 8.5.

$$\text{Report} = \left[\left[\left\langle ID_{pseudo}, \text{Event}\right\rangle\right]_{K_{pub}^{LE_{id}}} \middle\| ID_{inc}\right]_{ring_{sig}} \tag{8.4}$$

$$\text{Event} = \left[\left\langle E_{type}, E_{loc}, T_{min}, T_{max}\right\rangle\right] \tag{8.5}$$

8.5 Case Study for Framework Validation

In this section, we provide a detailed use case analyzing the need for extending crime investigations and associated digital forensics into the realm of crowd computing. The use case also serves as a good application for validating the proposed framework's usefulness and applicability. Numerous other crime investigations can benefit from the proposed solution, but due to chapter length restrictions, we only provide statistics of such major crimes, enabling readers to appreciate the gamut of applications possible.

8.5.1 Drive-By Shooting Hypothetical Case

8.5.1.1 Case Background

A drive-by shooting refers to an incident when someone fires a gun from a vehicle at another vehicle, a person, a structure, or another stationary object. Many drive-by shootings involve multiple suspects and multiple victims. Using a vehicle allows the shooter to approach the intended target without being noticed and then to speed away before anyone reacts. Deaths of innocent bystanders often receive significant media attention and result in passionate public outcry. This is often followed by overwhelming support from the community in sharing information and providing leads, in support of which a secure, anonymous reporting mechanism has to be established.

8.5.1.2 Crime Scene

The crime scenes of drive-by shootings often dissolve or dissipate rapidly as physical evidence is destroyed, witnesses leave the scene, and recollections of what occurred are influenced by discussions among witnesses and neighbors. Victims and other citizens who witness drive-by shootings are often reluctant to provide information to the police. This reluctance may stem from a fear of reprisal, from general community norms discouraging cooperation with the police, or from principles of personal retribution in the community. Therefore, some jurisdictions immediately deploy specially trained response teams to freeze the scene, preserve physical evidence, and ensure that witnesses remain present for questioning and are kept separate from one another. It is very important to collect as much information from potential witnesses as possible (Table 8.3).

LE's response alone is seldom effective in reducing or solving drive-by-shooting incidents. Therefore, LE agencies carefully consider who in the affected community are responsible citizens and care about the community and can help LE better respond to the incident. The responsibility of responding, in some cases, may need to be shifted toward those who have the capacity to implement more effective

Table 8.3 Information Gathered by First Responders Used on a Crowd-Assisted Web Portal

Drive-by Shootings	E_{type}	DbS
Incident location	E_{loc}	4th & Turner St.
Time window	$\langle T_{min}, T_{Max} \rangle$	$\langle 2.33\ AM, 2.47\ AM \rangle$

responses, and often will require cooperation from members of the community. The initial few hours after the incident provide the window with the highest opportunity for apprehending the perpetrators. Therefore, it is critical that a response strategy be decided and implemented as quickly as possible.

8.5.1.3 Anonymous Platform for Crowd Assistance

Let us assume that with the information gathered by the first responders and from 911 calls, the LE agency with jurisdiction of the case prepares the crowd-assisted platform template as follows.

After a drive-by shooting incident, assume that LE immediately posted a web portal for witnesses to submit information, in an anonymous manner. For discussion, let us assume that the citizens who either witnessed the incident or have any hearsay information uploaded it using the crowd-assisted web portal. Two sample reports are shown in Figure 8.3, and complete details of four such reports are presented in Table 8.4. With this information, LE can narrow down the following critical piece of information, which can be used to narrow down potential suspects and the getaway vehicle: dark sedan (possibly navy blue or black), vehicle had three occupants, vehicle has special paint job (likely a gang symbol), male driver wearing baseball cap, shooter in late teens with a hooded sweatshirt and possibly long hair, partial vehicle tag reads "HOT."

Subsequently, LE runs some queries on internal databases trying to identify anyone who fits the description; LE also runs the vehicle plate "HOT," looking for

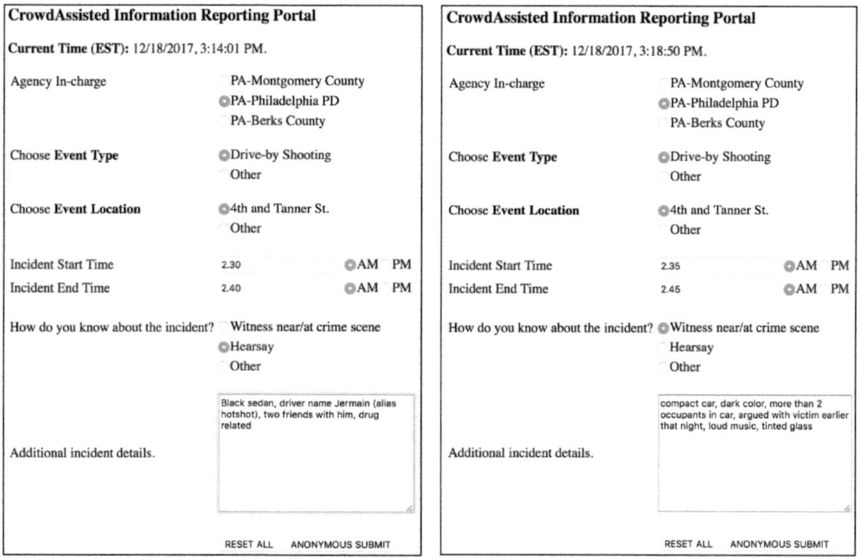

CrowdAssisted Information Reporting Portal

Current Time (EST): 12/18/2017, 3:14:01 PM.

Agency In-charge	PA-Montgomery County ● PA-Philadelphia PD PA-Berks County
Choose **Event Type**	● Drive-by Shooting Other
Choose **Event Location**	● 4th and Tanner St. Other
Incident Start Time	2.30 ● AM PM
Incident End Time	2.40 ● AM PM
How do you know about the incident?	Witness near/at crime scene ● Hearsay Other
Additional incident details.	Black sedan, driver name Jermain (alias hotshot), two friends with him, drug related

RESET ALL ANONYMOUS SUBMIT

CrowdAssisted Information Reporting Portal

Current Time (EST): 12/18/2017, 3:18:50 PM.

Agency In-charge	PA-Montgomery County ● PA-Philadelphia PD PA-Berks County
Choose **Event Type**	● Drive-by Shooting Other
Choose **Event Location**	● 4th and Tanner St. Other
Incident Start Time	2.35 ● AM PM
Incident End Time	2.45 ● AM PM
How do you know about the incident?	● Witness near/at crime scene Hearsay Other
Additional incident details.	compact car, dark color, more than 2 occupants in car, argued with victim earlier that night, loud music, tinted glass

RESET ALL ANONYMOUS SUBMIT

Figure 8.3 Crime report submitted by citizens using the crowd-assisted portal. (a) Hearsay report. (b) Witness report.

Table 8.4 Summary of Crowd-Assisted Data for the Drive-By Shooting Use Case

Report ID	Vehicle Info	Occupant(s) Info	Additional Info
DbS-121817-1518	Dark compact car	>2	Tinted glass, loud music, argued with victim earlier that night
DbS-121817-1514	Black sedan	3	Possibly Nissan; word on street is Jermain (alias hotshot) was driving with two friends; drug related
DbS-121817-0630	Navy blue coupe	3	Gang-related paint work, driver wearing hat, partial vehicle tag "HOT"; shooter in back seat, had long hair
DbS-121817-0420	Dark sedan	>3	Tinted windows, shooter wearing hooded sweatshirt, male driver, very loud music, gang-related paint job on vehicle; driver wearing baseball cap

a match. They do not come up with any definite subjects matching the suspect profile, but they locate a black Chevy Malibu with the tag "HOT SHOT," registered to a white male, that was reported as stolen. With all the above information, LE has enough leads to post emergency alerts on local TV, highways, mobile phones, and so forth, asking for any further leads.

8.6 Conclusion

Effective community policing initiatives and programs help to build trust between citizens and LE, which can help LE deal more effectively during times of crisis. More importantly, a healthy relationship between the community and LE helps mitigate angst and fear among community residents. This is fundamental to promoting community harmony and ensuring a sense of safety and security for all citizens. A systematically designed crowd sensing community policing framework has great potential toward helping LE develop knowledge of activities in communities and miscreant pupils.

Today's mobiles collect tons of valuable data, which if utilized effectively can assist LE in addressing a myriad of problems. While citizens have embraced crowd computing in extremely diverse domains, they have serious privacy concerns when it comes to information sharing with LE. In this chapter, we have emphasized the benefits that crowd-assisted information garnering brings to every-day crime investigations. We have presented the design and implementation details of a formal crowd-assisted community policing framework. The proposed framework is designed with provably secure cryptographic primitives that provide conditional anonymity. The ring signature allows a signer to leak secrets anonymously, without the risk of identity escrow. At the same time, the ring signature provides great flexibility: no group manager, no special setup, and the dynamics of group choice. On the other hand, the ring signature is vulnerable to malicious or irresponsible signers in some applications, because of its anonymity.

Finally, we have presented a detailed use case validating the feasibility and usefulness of the proposed framework in various everyday societal crime investigations. Note that this work can be easily adopted outside of crime investigation. Some of the broader applications can include anonymous survey for increased accuracy in responses, especially in surveys where participants are conscious of being judged. Such conditionally anonymous groups can also be provided for community discussions of proposed changes to community zoning, legislative bills, and the proposal of amendments and bylaws, among others.

References

1. Tingxin Yan, Matt Marzilli, Ryan Holmes, Deepak Ganesan, and Mark Corner. Mcrowd: A platform for mobile crowdsourcing. In *Proceedings of the 7th ACM Conference on Embedded Networked Sensor Systems*, pp. 347–348. New York: ACM, 2009.
2. Vamshi Ambati, Stephan Vogel, and Jaime G Carbonell. Active learning and crowd-sourcing for machine translation. Available at: https://www.cs.cmu.edu/~jgc/publication/PublicationPDF/Active_Learning_And_Crowd-Sourcing_For_Machine_Translation.pdf (last accessed: 30 June, 2018).
3. Michael J. Franklin, Donald Kossmann, Tim Kraska, Sukriti Ramesh, and Reynold Xin. Crowddb: Answering queries with crowdsourcing. In *Proceedings of the 2011 ACM SIGMOD International Conference on Management of Data*, pp. 61–72. New York: ACM, 2011.
4. Ali Dehghantanha and Katrin Franke. Privacy-respecting digital investigation. In *Privacy, Security and Trust (PST), 2014 Twelfth Annual International Conference on*, pp. 129–138. New York: IEEE, 2014.
5. Crowdsolve wants to turn amateurs into true detectives. Available at: http://www.engadget.com/2014/12/15/crowdsolve-detective/ (last accessed: 30 June, 2018).
6. Adam Bender, Jonathan Katz, and Ruggero Morselli. Ring signatures: Stronger definitions, and constructions without random oracles. In *Theory of Cryptography Conference*, vol. 6, pp. 60–79. Berlin: Springer, 2006.
7. Masayuki Abe, Miyako Ohkubo, and Koutarou Suzuki. 1-out-of-n signatures from a variety of keys. In *Proceedings of the 8th International Conference on the Theory and Application of Cryptology and Information Security: Advances in Cryptology*, pp. 415–432. Berlin: Springer, 2002.

8. Yevgeniy Dodis, Aggelos Kiayias, Antonio Nicolosi, and Victor Shoup. Anonymous identification in ad hoc groups. In *Lecture Notes in Computer Science*, vol. 3027, pp. 609–626. Berlin: Springer.

9. Javier Herranz and German Saez. Forking lemmas for ring signature schemes. In *International Conference on Cryptology in India*, pp. 266–279. Berlin: Springer, 2003.

10. Dan Boneh, Craig Gentry, Ben Lynn, and Hovav Shacham. Aggregate and verifiably encrypted signatures from bilinear maps. In *Lecture Notes in Computer Science*, vol. 2656, pp. 416–432. Berlin: Springer.

11. Xavier Boyen. Mesh signatures: How to leak a secret with unwitting and unwilling participants. In *EUROCRYPT07*, vol. 4515 of LNCS, pp. 210–227. Berlin: Springer, 2007.

12. Hovav Shacham and Brent Waters. *Efficient Ring Signatures Without Random Oracles*, pp. 166–180. Berlin: Springer, 2007.

13. Eiichiro Fujisaki and Koutarou Suzuki. Traceable ring signature. *IEICE Transactions on Fundamentals of Electronics, Communications and Computer Sciences*, vol. 91(1), pp. 83–93, 2008.

14. Shen Noether and Sarang Noether. Monero is not that mysterious. MRL-0003, Monero Research Lab. 2014. Available at: https://lab.getmonero.org/pubs/MRL-0003.pdf (last accessed: 30 June, 2018).

15. Mihir Bellare and Phillip Rogaway. Random oracles are practical: A paradigm for designing efficient protocols. In *Proceedings of the 1st ACM Conference on Computer and Communications Security*, pp. 62–73. New York: ACM, 1993.

16. Andreas Pfitzmann and Marit Köhntopp. Anonymity, unobservability, and pseudonymity: A proposal for terminology. In *Designing Privacy Enhancing Technologies*, pp. 1–9. Berlin: Springer, 2001.

17. Sarah Meiklejohn. An exploration of group and ring signatures. 2011. Available at: https://cseweb.ucsd.edu/~smeiklejohn/files/researchexam.pdf (last accessed: 30 June, 2018).

18. NHTSA/DOT. Traffic safety facts 2013 data—pedestrians.

19. Karl Kim, Pradip Pant, and Eric Yamashita. Hit-and-run crashes: Use of rough set analysis with logistic regression to capture critical attributes and determinants. *Transportation Research Record: Journal of the Transportation Research Board*, vol. 2083, pp. 114–121, 2008.

20. Kara E. MacLeod, Julia B. Griswold, Lindsay S. Arnold, and David R. Ragland. Factors associated with hit-and-run pedestrian fatalities and driver identification. *Accident Analysis & Prevention*, vol. 45, pp. 366–372, 2012.

21. National Crime Information Center. 2013 NCIC missing person and unidentified person statistics.

22. National Crime Information Center. 2014 NCIC missing person and unidentified person statistics.

23. National Crime Information Center. 2015 NCIC missing person and unidentified person statistics.

24. National Crime Information Center. 2016 NCIC missing person and unidentified person statistics.

25. Ronald Rivest, Adi Shamir, and Yael Tauman. How to leak a secret. In *Advances in Cryptology, ASIACRYPT 2001*, pp. 552–565. Berlin: Springer, 2001.

26. Yuichi Komano, Kazuo Ohta, Atsushi Shimbo, and Shin-ichi Kawamura. Toward the fair anonymous signatures: Deniable ring signatures. In *CT-RSA*, vol. 6, pp. 174–191. Berlin: Springer, 2006.

27. Shengke Zeng, Shaoquan Jiang, and Zhiguang Qin. A new conditionally anonymous ring signature. In *Computing and Combinatorics*, pp. 479–491. Berlin: Springer, 2011.
28. Anna Lysyanskaya. Conditional and revocable anonymity: An overview. 2011. Available at: https://pdfs.semanticscholar.org/presentation/f1a7/b96b537bbfe15d52e 4897307400a741135f4.pdf (last accessed: 30 June, 2018).
29. David Chaum and Eugene Van Heyst. Group signatures. In *Workshop on the Theory and Application of Cryptographic Techniques*, pp. 257–265. Berlin: Springer, 1991.

Chapter 9

Applicability of Lightweight Stream Cipher in Crowd Computing: A Detailed Survey and Analysis

Subhrajyoti Deb, Rohit Upadhya, and Bubu Bhuyan

Contents

9.1 Introduction

A new collaborative computing paradigm has evolved in recent times based on the strength of crowdsourcing, automation, and machine learning. Crowdsourcing is the combination of two terms, *crowd* and *outsourcing*, and refers to the execution of a task efficiently by distributing its subtasks to different groups. The idea of crowdsourcing was introduced in 2006 by Jeff Howe and Mark Robinson. A few examples of crowdsourcing are Wikipedia, Google Mapathon, threadless, and InnoCentive. Crowd computing has the advantage of efficiently solving several practical problems, such as image tagging and audio translation, which are very difficult for a computer alone to do. A generalized crowd computing framework has been discussed in [1,2]. So, it has applicability in wide variety of application domains, including communication, healthcare, vehicular communication, and home appliances. Due to human engagement, several security and privacy concerns are identified in crowd computing; a few of them are listed below.

■ Multiple mobile phones and other handheld devices are connected to different networks, and these devices may be regulated by malicious owners to launch attacks.
■ Private information needs to be safeguarded when a task is distributed to a subgroup in crowdsourcing.
■ The security of heterogeneous data produced by several low-end devices connected through different technologies like wireless sensor networks (WSNs), Bluetooth, radio frequency identification (RFID), and wireless mesh networks (WMNs) needs to be preserved.

This chapter presents security and privacy issues in the context of heterogeneous devices. Currently, the Internet of Things (IoT) intends to connect billions of smart and constrained objects to the Internet. In this regard, IoT will be an important part

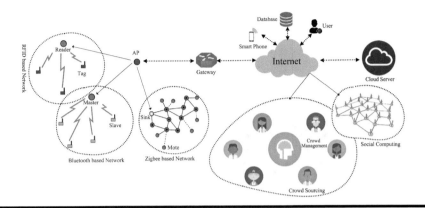

Figure 9.1   Structural view of crowd computing.

of crowd computing. All the connected objects are supposed to produce high volumes of data and transfer the data to the cloud server for further processing. The components are (1) devices (sensor nodes, route/relay nodes, actuator), (2) local network (gateway nodes), (3) the Internet, and (4) end-user devices (desktop, laptop, smartphone, etc.). Many of the heterogeneous devices, connected with different networks, are resource constrained in nature. The bulk amounts of data that are generated from these objects are in different forms and types. Multiple devices can join on the Internet through different technologies (like WSNs, Bluetooth, RFID, and WMNs) with the help of gateways. The gateway node connects two dissimilar networks. It can convert different types of data formats from various devices into a standard one and make them available to the remote user. Some of these gateways may require additional functions, such as local storage and a user interface. Local data can be sent to the remote cloud or server through the Internet. A remote user can give the command to the cloud through the Internet or access the processed sensor data for their use. A schematic diagram of a smart and constrained environment is shown in Figure 9.1.

It is very important to address the common technical specifications [3], such as security, scalability, dependability, quality of service, interoperability, and portability. Developing the cryptographic primitives for solving security issues in a constrained environment is still a live research problem. The cryptographic primitives should be lightweight as well having high throughput with a smaller hardware footprint, less power requirements, and so forth. Symmetric key cryptographic primitives have the advantage of higher throughput over asymmetric ones, and therefore symmetric key cryptographic primitives are widely used for bulk data encryption and decryption. Specifically, stream ciphers have the advantages of higher throughput, lower latency, and a smaller error propagation effect than block ciphers. The lightweight stream cipher is an effective security solution for low-end devices. Specifically, the linear feedback shift register (LFSR)–based stream cipher is characterized by its lightweight property and easy implementation in hardware. A few examples of popular stream ciphers are A5/1 used in global system for mobile communications (GSM) security, E0 used in Bluetooth, RC4 used in the Secure Sockets Layer (SSL), and Espresso used in the 5G wireless network. Currently, Manifavas et al. present a detailed survey on the lightweight stream cipher [4]. Our chapter presents a detailed analysis of the lightweight stream cipher.

The rest of the chapter is organized as follows: Section 9.2 provides the taxonomy of the lightweight stream cipher. Section 9.3 provides the randomness test analysis of the stream cipher. Section 9.4 provides an overview of common lightweight stream cipher theoretical attacks. In Section 9.5, we discuss some security-related research issues. Finally, Section 9.6 concludes the chapter.

9.2 Taxonomy of Stream Cipher

This section discusses the taxonomy, characteristics, advantages, and cryptographic features of the stream cipher. Before starting the stream cipher, it needs to address the basic terms of cryptology. The word *cryptology* is the combination of two Greek words, *kryptós* (hidden) and *logia* (study) [5]. Cryptography is the study of mathematical techniques used for achieving secured communication over an unsecured channel. Concurrently, cryptanalysis refers to finding weaknesses in cryptosystems. Cryptology comprises cryptography and cryptanalysis. More specifically, cryptology is the study and practice of hiding information, and it is the art of science and information security.

Cryptographic primitives are designed to deal with basic security issues, like confidentiality, integrity, authentication, and nonrepudiation. It can be classified into two categories: symmetric key and asymmetric key primitives. Symmetric key ciphers are again subdivided into two categories: block cipher and stream cipher. A stream cipher can generate cryptographically secure pseudorandom sequences. A pseudorandom number is highly preferable for encryption and decryption in cloud computing, WSNs, communication channels, and so on. A few prominent lightweight stream ciphers are Grain, WG, Trivium, SNOW, Salsa 20, Sprout, Espresso, Lizard, Fruit, and Plantlet.

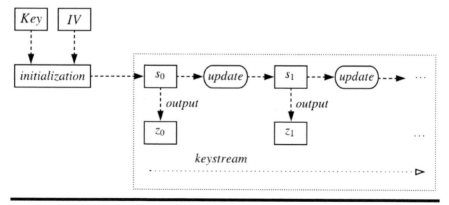

Figure 9.2 A schematic diagram of a simple stream cipher.

A lightweight stream cipher primarily consists of two components: key scheduling algorithm (KSA) and pseudorandom generation algorithm (PRGA). KSA increases the randomness properties inside the internal state of the stream cipher. PRGA always updates the internal state at every phase and after each update phase produces a key-stream bit. A schematic diagram of the stream cipher keystream generation process is shown in Figure 9.2. A stream cipher is known as synchronous if the keystream is fully dependent on the secret key; otherwise, it is known as an asynchronous stream cipher. Usually, synchronous stream ciphers do not propagate any error transmission.

In a mathematical way, the lightweight stream cipher can be defined by two functions, *F*, *G*, and internal state σ (finite state vector). Here, σ is updated at every phase by the function of *F*. Specifically, σ is the outcome of the state vector at time *t*, and *K* denotes the secret key and *P* represents the plaintext of the cipher [6].

$$\sigma_{t+1} = F\left(\sigma_t, p, k\right)$$

$$K_t = F\left(\sigma_t, p\right)$$

Crowd computing provides various sharing and scalable services as required from any location. Thus, the cryptographic algorithm is very much important for security assurance. Due to high computational complexity, lightweight cryptographic algorithms are necessary for low-end devices. In order to enhance security, the

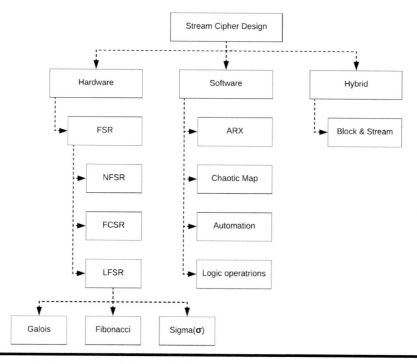

Figure 9.3 Taxonomy of stream ciphers.

lightweight stream cipher is a computationally efficient cryptographic algorithm, and it can be used by different applications. Stream cipher design schemes can be classified into three categories. Figure 9.3 shows the taxonomy of lightweight stream ciphers. All the parts have been briefly classified in this section.

9.2.1 Hardware Implementation

In crowd computing, high-speed computation is very important for heterogeneous devices. Thus, a feedback shift register (FSR) is commonly used for pseudorandom sequence generation and fast computation. Hardware-based FSR always maintains a good trade-off between security and efficiency. Particularly, sensor-based networks require a pseudorandom generator for secret key establishment and data transfer. In this section, we review hardware-based components that are generally used in the modern stream cipher.

9.2.1.1 Feedback Shift Register

An FSR can be considered to be a cascade of flip-flops that shift the bit array stored by one position after each pulse. When cells are initialized by words (or bits), this is known as a secret value or seed. Usually, the feedback function describes how the values of cells are computed, and this function can be represented as linear or nonlinear.

9.2.1.1.1 Linear Feedback Shift Register

An FSR is known as an LFSR if its feedback function is linear. An LFSR is basically the hardware or software realization of a finite state machine. Flip-flops are used to implement a standard LFSR. Usually D-flip-flops are used to represent different states of the LFSR. A practical implementation of LFSR can easily generate

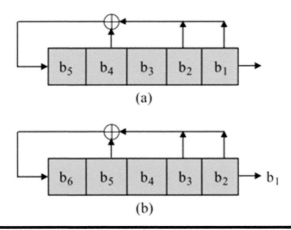

Figure 9.4 (a) Structure of LFSR. (b) First output stage.

pseudorandom numbers, but LFSR is not cryptographically secure [5]. The period of the sequence produced by this LFSR will be $(2^d - 1)$. The two-stage operation of an LFSR is shown in Figure 9.4.

We know that any shift register is controlled by an external clock. For instance, at each time interval, flip-flop content is shifted one stage to the right. Here, the value of the rightmost stage is b_1, which is the first output of LFSR. The new value of the leftmost stage of the feedback bit is obtained by the linear relation of the contents of the register. For an L-bit LFSR, the recursion relation can be written as

$$b_i(t+1) = b_{i+1}(t)$$

$$b_L(t+1) = \overset{L}{\underset{i=1}{\oplus}} c_i b_i(t)$$

where c_i depends on the feedback logic of the LFSR. It should be noted that XOR (modulo 2 operations) is used for linear operations, whereas AND, OR gates are used for nonlinear operations.

9.2.1.1.1.1 Fibonacci LFSR

Fibonacci LFSR logic is involved in the feedback path. In this LFSR, data flow is from left to right and the feedback path is from right to left. A max-length LFSR cycles through $2^n - 1$ states. However, it will never change if it contains all 0s.

Let us describe the above concepts using the following equation:

$$f = x^{16} + x^5 + x^3 + x^2 + 1$$

The power of the x values corresponds to the tap positions. However, here 1 does not correspond to any tap. It only corresponds to the input for the first bit.

Figure 9.5 describes the design of the LFSR in accordance with the equation. The following are the conditions for the LFSR to be of maximum length:

1. The total number of taps should be even.
2. The taps should be relatively prime to each other.

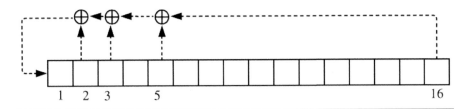

Figure 9.5 A structure of a Fibonacci LFSR.

However, these conditions are *necessary* but *not sufficient* for a cryptographically secure bitstream.

9.2.1.1.1.2 Galois LFSR   Although the design of the Galois LFSR varies quite a bit from that of the Fibonacci LFSR, for the same tap polynomial it produces the same result as the Fibonacci LFSR. The only difference is that the Galois LFSR is in the opposite orientation as the Fibonacci LFSR. In Galois LFSR, data flow is from left to right and the feedback path is from right to left [7]. Meanwhile, the input to the last position will directly be the output of the first position. Figure 9.6 shows the design of the Galois LFSR for the equation that was used in the Fibonacci LFSR.

9.2.1.1.1.3 Sigma (σ)-LFSR   An LFSR can be used as a linear recurrence relation of finite order over the field F_q. In the case of a binary, q is 2. In order to achieve good cryptographic properties, one of the main requirements is a maximum period. Currently, Zeng et al. have introduced a generalization of word-oriented LFSR known as σ-LFSR [8]. Moreover, it increases the Hamming weight of the characteristic polynomial. For instance, m is a positive integer of the field F_{q^m} with q^m elements and the vector space F_q^m of dimension m over F_q. σ-LFSR use less memory and computer instructions. Word-oriented LFSRs are used for fast software implementation and also have been fruitful for many cryptographic schemes
Figure 9.7 describes the basic design of the σ-LFSR structure.

9.2.1.1.2 Nonlinear Feedback Shift Registers

An FSR is known as a nonlinear feedback shift register (NLFSR) if its feedback function is nonlinear. An NLFSR uses a nonlinear function to update the

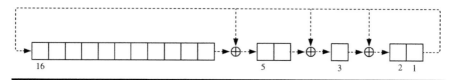

Figure 9.6   A structure of a Galois LFSR.

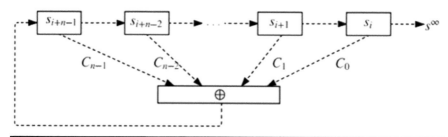

Figure 9.7   A schematic diagram of σ-LFSR over F_{q^m}.

internal state of the shift register. For an n-bit shift register p, the next state is defined as

$$p_{i+1}\left(s_0, s_1, \cdots, s_{n-1}\right) = p_i\left(s_1, s_2, \cdots, f\left(s_0, s_1, \cdots, s_{n-1}\right)\right)$$

p is the nonlinear feedback function and s is the state.

NLFSRs are major components in many modern-day stream ciphers, like Grain and Fruit. Generally, NLFSRs are known to have a better chance of avoiding the cryptanalytic attack than LFSRs. NLFSRs utilize multiplication (OR, AND operation). It is seen from the equation that there exists at least one term that is actually a multiplication of two or more terms. The polynomial f will have the following properties [9]:

1. For an NLFSR having n stages, its f will have a degree of 2^n.
2. There exists a one-to-one correspondence between the cycles of the NLFSR and the irreducible factors of f.
3. There exists a mapping of the roots of f to the output sequence of the register.

Figure 9.8 describes the structure of NLFSR.

9.2.1.1.2.1 Advantages of LFSR and NLFSR in Crowd Computing

1. A group of people are treated as end users who acquire or pay the crowdsourcing services at specific costs. Moreover, the end users require different types of hardware-based devices for basic communication. In that case, LFSR or NLFSR is one of the good options for security. As per requirements, end users can easily perform encryption and decryption tasks in short time duration.

In the next part of the chapter, we separately review the common properties that are used in the construction of keystream generators. Here, we study the common security components that are used to judge the cipher nonlinear functions. Here, all the necessary mathematical preliminaries are presented—all the mathematical preliminaries related to LFSR- and NLFSR-based stream cipher design.

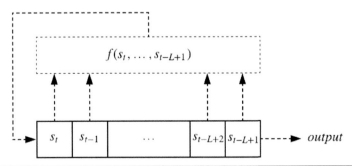

Figure 9.8   A schematic diagram of an NLFSR.

■ **Boolean function:** A Boolean function is a mapping function from $F_2^n \rightarrow F_2$, over the finite field, with two elements {0, 1}. If the number of combination mapping consists of an equal number of 1s and 0s, then the Boolean function is known as balanced.

■ **Algebraic normal form:** Usually, every Boolean function has a unique representation as a multivariate over F_2, which is known as the algebraic normal form (ANF). This function can be represented as

$$f\left(x_1, x_2 \ldots x_n\right) = c_0 \oplus \sum_{1 \le i \le n} c_i x_i \oplus \sum_{1 \le i \le j \le n} c_i x_i x_j \oplus c_{(1\ldots n)} x_1, \ldots x_n$$

where the coefficients c_0, c_i, $c_{i,j}$, ..., $c_{(1\ldots n)} \in F_2$. In this function, the number of variables in the highest-order product term (with coefficient nonzero) is known as the algebraic degree. In general, when the degree of the function f is at most 1, it can be described as an affine function. The affine functions with ($c_0 = 0$) are known as linear functions [10–12].

■ **Walsh transform:** This transformation function is an n-variable Boolean function. In that case, $c = \{c_1 \ldots c_n\} \in F_2^n$ and the n-variable linear function can be represented as $l_c(x) = c_1 x_1 \oplus \ldots \oplus c_n x_n$. So, this transformation function can be described as

$$W_f\left(c\right) = \sum_{x \in F_2^n} (-1)^{f_x \oplus l_c(x)}$$

From the above definition of $W_f(c)$, it can be observed that when $f_x \oplus l_c(x)$ is 0, the sum is increased by 1, and when this value is 1, the sum is decreased by 1.

■ **Nonlinearity:** The nonlinearity of a Boolean function f of n variables can be described as the distance between the function and the set of all possible affine functions. Nonlinearity can be defined in terms of the Walsh transform as given below.

$$nl\left(f\right) = 2^{n-1} - \frac{1}{2} \max \left|W_f\left(c\right)\right|$$

■ **Correlation immunity:** A Boolean function f on F_2^n is said to be a correlation of order m, where ($1 \le m \le n$) if the output of f and any m input variables are statistically independent [11,12]. A Boolean function f on F_2^n is correlation immune of order m if and only if $W_f(c) = 0$ for all vectors such that $c \in F_2^n$ such that $0 \le |(c)| \le m$.

Here, different variable Boolean functions are constructed on a set of Boolean variables. All the experiments are done with the SageMath tool.* Table 9.1 shows the typical values of parameters such as balancedness, nonlinearity, maximum nonlinearity, algebraic immunity, correlation immunity, autocorrelation, and Walsh transform.

We know that large sequences of a binary string should satisfy good randomness characteristics, such as maximum period, balancedness, perfect run distribution, and good autocorrelation. But their main shortcoming is low linear complexity, and it leads to the shortest LFSR length for an assigned binary output sequence. The Berlekamp–Massey algorithm is used to evaluate the linear complexity of a sequence [13]. Thus, three types of common generators, like nonlinear combiners, nonlinear filter generators, and clock-controlled generators, are used in the keystream construction.

9.2.1.1.2.2 Filter Generator
In general, filter generators work on a single LFSR. Typically, the internal state of the LFSR is clocked linearly. Some chosen stages from the LFSR are used as the input to the nonlinear function that is applied to the keystream. Here, n-length LFSR is used (stages like a_1, a_2, …, a_r, …, a_n), along with a nonlinear filter function f of order m, where ($m < n$). It must be ensured that for the selected inputs, the keystream is generated by strong linear complexity [4,9]. Figure 9.9 shows the general structure of an LFSR-based filter generator.

9.2.1.1.2.3 Combination Generators
Combination generators use a sequential combination of LFSRs. Here, the final output is produced by a nonlinear Boolean function or the combining function. A mapping of the input is done by the nonlinear function to produce the output, which is a binary variable. In order to achieve maximum length, LFSR should be selected carefully where lengths are co-prime. In that case, the feedback function must be primitive with the distinct degree [9]. The function f should satisfy a higher order of correlation immunity. Figure 9.10 shows the structure of LFSR-based combination generator.

9.2.1.1.2.4 Clock-Controlled Generators
These generators are used to overcome the linearity of the LFSR output by applying irregular clocking. They use numerous registers for a nonlinear output sequence. The registers' state depends on external or internal clocking [9]. The register that controls the clocking is known as the clocking register (CR). Moreover, the generator register (GR) controls the keystream (Figure 9.11).

---

* SageMath is a free open-source mathematics software system. Available at: http://www.sagemath.org/

Table 9.1 Boolean Function with Its Cryptographic Properties

Boolean Function	Balancedness	Nonlinearity	Algebraic Immunity	Correlation Immunity	Autocorrelation	Walsh Transform
$x_3 + x_2x_3 + x_1x_2x_3$ (3-variable Boolean polynomial ring)	True	2	2	0	(8, 0, 0, 0, 0, 0, -8, 0)	(0, 0, 4, 4, -4, 4, 0, 0)
$x_0x_1 + x_2 + x_3$ (4-variable Boolean polynomial ring)	False	3	2	0	(16, -4, -4, 4, -12, 4, 4, -4, -12, 4, 4, -4, 12, -4, -4, 4)	(2, 2, 2, 2, 2, 2, 2, 2, 2, 2, 2, -6, -6, -6, 10)
$x_0x_1 + x_1x_2 + x_2x_3 + x_3x_4 + x_4x_5$ (6-variable Boolean polynomial ring)	False	28	2	0	(64, 0, 0, 0, 0, 0, 0, 0, 0, 0, 0, 0, 0, 0, 0, 0, 0, 0, 0, 0, 0, 0, 0, 0, 0, 0, 0, 0, 0, 0, 0, 0, 0, 0, 0, 0, 0, 0, 0, 0, 0, 0, 0, 0, 0, 0, 0, 0, 0, 0, 0, 0, 0, 0, 0, 0, 0, 0, 0, 0, 0, 0, 0, 0)	(-8, -8, -8, -8, 8, -8, -8, -8, -8, 8, 8, -8, -8, -8, -8, 8, -8, -8, 8, 8, -8, 8, 8, -8, -8, 8, 8, -8, 8, 8, 8, 8, -8, -8, 8, 8, -8, 8, 8, 8, -8, 8, 8, -8, 8, 8, 8, 8, -8, 8, 8, 8, -8, 8, 8, 8, 8, -8, 8, 8, 8, -8)

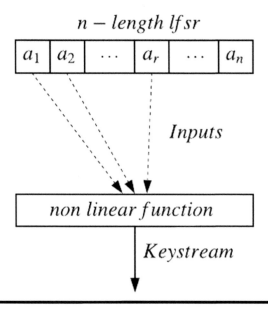

Figure 9.9   LFSR-based filter generator.

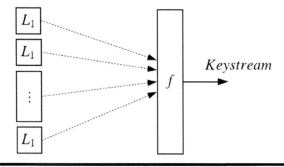

Figure 9.10   LFSR-based combination generator.

9.2.1.1.3 Feedback Carry Shift Register

A feedback carry shift register (FCSR) is nothing but a shift register that has the property of memory. The FCSR is associated with an odd positive integer $p \in Z$, also known as a connection integer. The feedback to the shift register is given by the 2-adic* digits of $p + 1$.

$$p + 1 = p_1 2 + p_2 2^2 + \cdots + p_r 2^r$$

---

* http://mathworld.wolfram.com/p-adicNorm.html

Figure 9.11 LFSR-based clock-controlled generator.

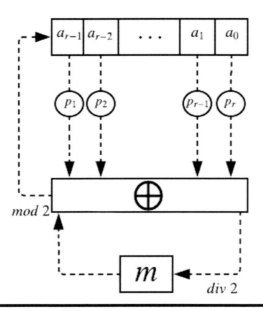

Figure 9.12 A schematic diagram of an FCSR over F_{2^m} .

There are $r = \lceil log(p+1) \rceil$ stages used by the shift register and also a maximum of $\lceil log(r) \rceil$ additional bits of memory. $[a_{r-1}, a_{r-2}, \cdots, a_0]$ specifies the state of the register [9].

Figure 9.12 represents the diagram of an FCSR.

9.2.2 Software Implementation

The stream cipher is used as it is fast and easy to implement in both hardware and software. Some basic specifications are discussed below.

9.2.2.1 ARX

ARX ciphers use modular add, rotate, and XOR operations. A large number of symmetric primitives use modular ARX, and also it has been used in the cryptographic field for the last 20 years [4]. ARX is used in different block and stream ciphers, like Skein, BLAKE, CubeHash, and Salsa20.

The main advantages of ARX are

- Ease of implementation in hardware and fast computation.
- A timing attack is not possible.
- Rotational cryptanalysis is very efficient.

A few disadvantages of ARX are

- Linear and differential cryptanalysis are possible.
- A side-channel attack is also possible.

9.2.2.2 Chaotic Maps

In 1989, Matthews presented the first idea of a stream cipher based on a one-dimensional chaotic map. Chaotic maps have many important properties in cryptography. They can be used in different ciphers for the purpose of confusion, diffusion, balance, avalanche property, and so forth. After use, the cipher outcome behavior is random with good statistical properties.

Their main advantages are listed below.

- System complexity is high due to high dimensionality and chaos.
- Hardware implementation is also possible.
- A discrete chaotic cryptosystem is used in different applications.

9.2.2.3 Automation

Cellular automata (CA) are special forms of finite state machines that can be examined by the common cryptanalytic methods. They can be interpreted as arrays of identical cells in an n-dimensional space and identified by various parameters. Moreover, linear CA produce randomness features through the employment of BIST schemes, self-checking, and so forth. Any kind of LFSR-based stream cipher or clock-controlled shrinking generator can be investigated and classified with a subset of linear CA [14].

9.2.2.4 Logic Operations

AND, OR, and XOR are the basic logical operations of a stream cipher. For example, the XOR operation is used when executing one-time passwords (OTPs) on computers. In 1949, Claude E. Shannon mathematically proved that OTP is perfectly secure for encryption. Even though the OTP system is secure, has a great flaw that causes it to be inefficient: the key length is required to be as long as the total length of the plaintext. In modern communication, stream ciphers are always used for faster and less code compared with block ciphers. All the logical operations are provided by low-end devices, like bitwise operation or arithmetic operation.

9.2.3 Hybrid Implementation

Hybrid systems connect the two methods and utilize the beneficial features of both hardware and software. The hybrid concept mainly focuses on communications like RFID tags. This method actually comes from Rueppel's definition: "Block ciphers operate on data with a fixed transformation on large blocks of plaintext data; stream ciphers operate with a time varying transformation on individual plaintext digits."* Hummingbird and Eris are examples of a hybrid structure that is built up by block and stream ciphers.

We know that embedded computer systems involve our daily life in various forms. Currently, embedded systems play a major role in modern society due to their computing systems. In this context, the main aim is the reduction of the logic gates, which requires a lightweight cipher. This metric is known as gate equivalent (GE). In general, a small GE has a lower cost and consumes less power. In symmetric cipher hardware, realizations like GE, throughput, and technology values (μm) are tabulated (Table 9.2).

9.3 Randomness Test Analysis

Crowd computing applies a number of computing ideas and technologies for service orientations such as cloud-as-a-service and infrastructure-as-a-service. Security is also treated as a service. Similarly, data security is also considered a service. Data security is one of the major issues in cloud-based computing. Current block cipher– and stream cipher–based cryptographic algorithms play a vital role in securing data communications. The randomness of the cipher is a good indicator of its security evaluation [27]. In this part, the randomness checking of the keystream generated by a stream cipher is one of the prominent indicators for resisting many known cryptanalytic attacks.

Recently, packages of statistical test suites have been invented to measure the cryptographic security properties of ciphers to investigate their randomness. Currently, a number of statistical tests for pseudorandom number generators (PRNGs) are developed, but huge interest has grown among the research communities for Dieharder [29], the NIST test suite, and TestU01. In January 2016, the National Institute of Standards and Technology (NIST) of the United States, formerly known as the National Bureau of Standards, published an updated version of Special Publication (SP) 800-90B (second draft) to examine the validation of entropy sources for the standardization of the deterministic random bit generator (DRBG). NIST SP 800-90 is a series of three documents considered for generating random numbers for different security applications [28]. The SP 800-90 series is basically a two-way phase: first, the entropy source behavior is inaccessible for

---

* https://pdfs.semanticscholar.org/2b8d/00f8051b835b89781a59726af0647f972fdd.pdf

Table 9.2 Specification of Lightweight Cipher Hardware Realization

Stream Cipher Name	Key Size	IV Size	Internal State	Throughput	Area Size (GE)	Platform	Year	Reference
Sprout	80	70	320	100	813	0.18 μm	2015	[15]
Grain v1	80	64	160	100	1294	0.13 μm	2007	[16]
Fruit	80	70	130	100	918	0.13 μm	2016	[16]
Espresso	128	96	256	100	1497	90 nm	2015	[17]
Trivium	80	80	288	100	2580	0.13 μm	2004	[16]
Mickey 80 v2	80	$0 \leq IV \leq 128$	200	100	3188	0.13 μm	2005	[16]
F-FCSR-H	80	80	160	100	4760	0.13 μm	2008	[18]
Hummingbird	256	64	80	100	2159	0.18 μm	2010	[19]
Decim 80	80	64	192	100	2603	0.13 μm	2005	[18]
WG	80	80	319	100	1786	0.65 μm	2013	[20]
Edon 80	80	64	320	100	4969	0.13 μm	2005	[18]
Enocoro	80,128	64	176	100	2700	0.18 μm	2008	[21]

(Continued)

Table 9.2 (Continued) Specification of Lightweight Cipher Hardware Realization

Block Cipher Name	Key Size	Block Size	Structure, Rounds	Throughput	Area Size (GE)	Platform	Year	Reference
AES	128	128	SPN, 10	12.4	3400	0.13 μm	1998	[22]
DESL	184	64	Feistel, 16	44.44	2168	0.18 μm	2007	[22]
PRESENT 80	80	64	SPN, 31	200	1570	0.18 μm	2007	[23]
PRESENT 80	128	64	SPN, 31	11.4	1000	0.35 μm	2007	[23]
HIGHT	128	64	188, 2	188.2	3048	0.25 μm	2006	[24]
PRINCE	128	64	533.3	533.3	3491	0.13 μm	2012	[25]
SIMECK	128	64	133.3	133.3	1484	0.13 μm	2015	[26]

guessing and determining the seed value, and second, DRBG provides a large bit sequence that may be used to a great extent. In each test, p-value is calculated. If the p-value is greater than 0.01 for all the tests, the particular test is considered to have failed the randomness test. Few weaknesses have been found in the Sprout design. In this part, we focus on Sprout keystream randomness result. For the Sprout cipher, 10^6 random keystream bits have been generated (with 100 independent files using different keys and initialization vectors [IVs]) for randomness test analysis.

9.3.1 Description of Sprout

The main aim of the author of Sprout is to reduce the internal state in lightweight stream cipher. In Sprout, a state update function is used that depends on a secret key. The IV and key size of the cipher is 80 bits. Like the Grain 128a stream cipher, the Sprout keystream generator comprises two FSRs: one LFSR and one NLFSR. Here, an initialization function and an update function are both key dependent. The output function generates the keystream of the cipher. Sprout's maximal keystream range is 2^{40}, which can be constructed by the same IV. Sprout's cipher state size is 80 bits, and the structure of this stream cipher is like that of the Grain family. But this cipher has been broken by several cryptanalysis methods. The main interesting feature of the structure is that for the first time it uses the secret key bits during the PRGA. One may note that Sprout uses significantly less area than other existing lightweight stream ciphers. A schematic diagram of the Sprout cipher is shown in Figure 9.13. Here, we describe a few notations of the cipher.

t = clock-cycle number
$L^t = (l^t, l^{t+1}, \ldots, l^{t+39})$, state of the LFSR in clock t
$N^t = (n^t, n^{t+1}, \ldots, n^{t+39})$, state of the NLFSR in clock t
C^t = round constant at clock t (counter)
$K = (k_0, \ldots, k_{79})$, key
$IV = (iv_0, \ldots, iv_{69})$, initialization vector
k_t^* = round key bit generated during clock-cycle t
z_t = keystream bit generated during the clock-cycle t

Dieharder package consists of 31 fully independent statistical tests, as listed in Table 9.3. In fact, producing good-quality random numbers from the deterministic system is a live research problem. Indeed, it is very hard to generate truly random numbers from any deterministic machine. In our experiment, maximum tests are passed, except a few tests that are considered to exhibit high randomness of the cipher.

The p-value of a randomness test is constructed by the properties of statistical interpretation. The statistical interpretation outcomes provided by the test suite are more or less problem-solving in nature. Moreover, any single test is formulated by the number of samples, number of tuples, tsamples and psamples,

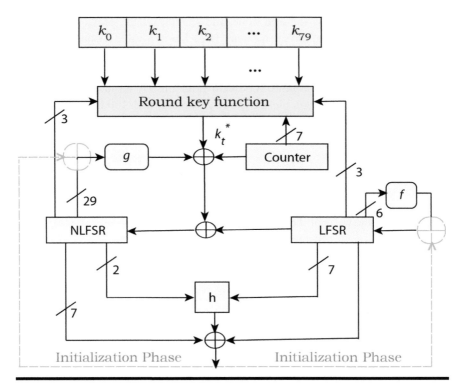

Figure 9.13 Structure of the Sprout stream cipher.

and so on. In the randomness test, we preferred to use the significance level of $\alpha = 0.01$, which is the default value for the test suite package. Using the Dieharder package, we analyze randomness results of binary sequences (keystream) generated by Sprout cipher. Specifically, for a few tests of Dieharder, several interpreted p-values are available. For the evaluation of randomness, we plotted the p-values along the x-axis and the tuple values along the y-axis. We also consider the several p-values that are generated for the sts serial, rgb lagged sum, rgb bitdist, rgb minimum distance, and rgb permutation tests. All the aforementioned tests are precisely described graphically: the sts serial results are in subplot (a); the rgb lagged sum test results in are subplot (b); the rgb bitdist, rgb minimum distance, and rgb permutation results are in subplot (c); and the remaining tests whose evaluation (assessment) resulted in one p-value are in subplot (d) of the Sprout cipher in Figure 9.14. Only a few tests show weak assessments, and all the results are shown in Table 9.4. In summary, Sprout cipher keystreams have passed the Dieharder randomness test, and only a few tests show weak assessments.

Table 9.3   Dieharder All-Tests Specification

Name of the Test	Ntuple	Remarks
Diehard Birthday Test	0	**psamples:** It represents the number of *p*-value samples per test.
Diehard OPERM5 Test	0	
Diehard 32x32 Binary Rank Test	0	
Diehard 6x8 Binary Rank Test	0	**tsamples:** It represents the number of trials used in each test.
Diehard Bitstream Test	0	
Diehard OPSO Test	0	**xo:** It represents the maximum number of psamples to decide the test is good.
Diehard OQSO Test	0	
Diehard DNA Test	0	
Diehard Count the 1s (stream) Test	0	
Diehard Count the 1s Test (byte)	0	**ntuple:** It fixes the ntuple size for tests on short bit strings that allow the length to differ.
Diehard Parking Lot Test	0	
Diehard Minimum Distance (2d Circle) Test	0	
Diehard 3d Sphere (Minimum Distance) Test	0	
Diehard Squeeze Test	0	
Diehard Sums Test	0	
Diehard Runs Test	0	
Diehard Craps Test	0	
Marsaglia and Tsang GCD Test	0	
STS Monobit Test	0	
STS Runs Test	0	
STS Serial Test (Generalized)	0	
RGB Bit Distribution Test	1–12	
RGB Generalized Minimum Distance Test	2–5	
RGB Permutations Test	2–5	
RGB Lagged Sum Test	0–32	
RGB Kolmogorov-Smirnov Test	0	

(Continued)

Table 9.3 (Continued)   Dieharder All-Tests Specification

Name of the Test	Ntuple	Remarks
Byte Distribution	0	
DAB DCT	0	
DAB Fill Tree Test	0	
DAB Fill Tree 2 Test	0	
DAB Monobit 2 Test	0	

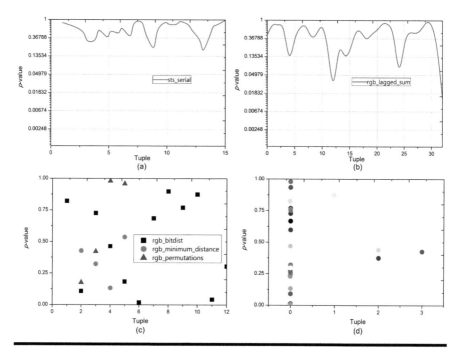

Figure 9.14   Dieharder randomness result of the Sprout cipher.

9.4 Attacks

Many stream ciphers have been developed over the years. The main advantages of the stream ciphers are fast computation, no fixed block size, and no propagation of errors. The current lightweight stream cipher design paradigm is very interesting; however, the success of a stream cipher design depends on its ability to resist cryptanalytic attacks. In this part, we discuss a few common theoretical attacks on stream ciphers. Here, we will present only seven attacks, but there are many others.

Table 9.4 Weak Randomness Test Results on Sprout

Dieharder version 3.31.1, Copyright Robert G. Brown						
Cipher Name	Test Name	Ntuple	Tsamples	Psamples	p-Value	Assessment
Sprout	sts_serial	11	100,000	100	0.99700233	Weak
	rgb lagged sum	29	1,000,000	100	0.99685544	Weak

9.4.1 Time–Memory Trade-Off Attack

The time–memory trade-off (TMTO) attack was first Introduced by Hellman in 1980 [30]. The concept behind the cryptanalytic TMTO attack is to efficiently invert an arbitrary one-way function. This approach is a middle way between two extremes using brute force and recording all the precomputed values in a table. There are two stages of a TMTO attack:

1. Precomputation phase
2. Online phase

9.4.1.1 Precomputation Phase

In the precomputation phase, the one performing the attack has to randomly pick m starting points $S_1, S_2, \cdots, S_m \in \{1, 2, \cdots, N\}$. For every $1 < i < m$, a chain has to be created by t successive applications of the function from the starting point S_1. In that case, an $m \times t$ Hellman's table is generated. Figure 9.15 depicts the Hellman table. The values in the rightmost column are called the end points (EP). Also, in order to reduce the excess consumption of space, only the starting points (S_i) and end points (EP_i) are stored after sorting them in increasing order of end points.

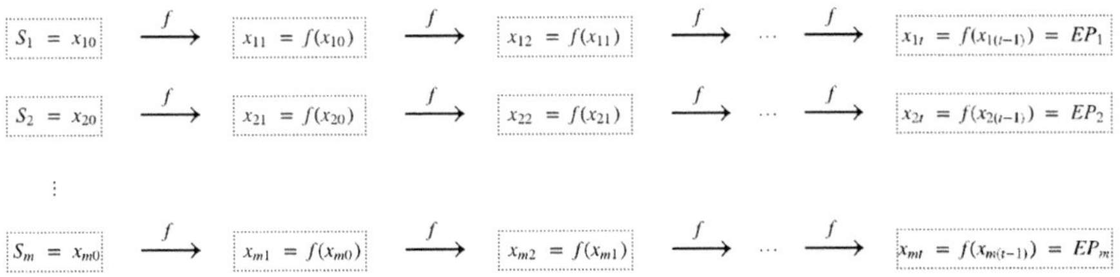

Figure 9.15 Hellman's table.

9.4.1.2 Online Phase

The basic objective of the attacker is to obtain the preimage of the value $x_0 = f(k)$ (k is the key). The attacker checks whether x_0 matches any value of EP_i. This can be done in logarithmic time as the EPs are already sorted. If the attacker is successful in finding an EP_i that is equal to x_0, then the value before the EP_i (or the preimage of the EP_i) is the required key.

9.4.2 Guess and Determine Attack

In this type of attack, linear cryptanalytic techniques are used to obtain the internal state of the cipher. In this process, an attacker is required to prepare some guess about how the internal state was formed [31]. After that, an attacker computes the actual register phases and operates under these assumptions to determine the values of the rest of the phases. Using the modified internal state, the attacker prepares the keystream and compares it with a recognized keystream. If the results are similar, then the attacker is capable of recovering the actual internal state.

9.4.3 Distinguishing Attack

We know that stream cipher generators are capable of producing a pseudorandom bitstream and it is indistinguishable from a truly random bitstream from the computationally bounded adversary. This is considered the keystream of the stream cipher random keystream bits that might be observed. Thus, the distinguishing attack is computed by the number of keystream bits that require examination for the expected event to be different from the structure-dependent (or initial state-dependent) patterns and keystream. In this attack, the main approach is to find the typical measures of statistical weakness that can pose a generic threat against any stream cipher construction. If any test indicates or specifies that the keystream generator has some biasness, this may lead to a distinguishing attack [9,31].

9.4.4 Algebraic Attack

This is one type of known plaintext attack, which assumes that the attacker has some sense or idea of a few plaintext bits and the corresponding ciphertext bits. Here, the main target is to obtain the "key" of the ciphers by exploring the set of polynomial equations. This attack was first introduced in stream ciphers by Courtois [9,31] and Courtois and Meier. During a strong algebraic attack, three stages are required to satisfy it:

1. Precomputation: In this stage, equations correlate key or internal state bits of the cipher.
2. Substitution and reduction: A set of equations is obtained and then explained to produce the key.

3. Solving: In this stage, the entire system of equations is determined by some linearization algorithm.

9.4.5 Correlation Attack

The correlation attack was introduced by Siegenthaler et. al [32], and its main intention is to obtain the initial state of the stream ciphers using a divide-and-conquer strategy on keystream generators with various parts, such as the Boolean function (nonlinear combination function). Moreover, the fundamental objective of this attack is to find the cipher's weak Boolean functions. Correlation is the connection of the binary bitstream, and flaws imply the closeness with which they correspond in behavior and appearance. For any stream cipher, the keystream sequence is determined as a noisy version of the output series of the LFSR sequence [11].

9.4.6 Cube Attack

In EUROCRYPT 2009, Dinur and Shamir proposed the cube attack, which is similar to the algebraic attack [33]. This attack is a method of chosen plaintext attack. Here, the output of a cryptosystem can be described as a polynomial of public variables and keys. Specifically, this attack can be tested on all cryptosystems if it is expressed as tweakable polynomials.

9.4.7 Fault Attack

Current lightweight stream cipher fault models normally depend on specific assumptions and consider a version of the cipher that is relatively weaker than the original cipher. Then the attacker should have the knowledge or some partial control in terms of state bit place and timing of fault injections. An attacker is able to reset the structure with the same key as many times as he wants through a guessing approach. A fault attack is basically divided into two parts: hard fault attack and soft fault attack. An attacker has the knowledge to guess and inject the fault at any random position of the state; this process is known as soft fault attack. In contrast, in a hard fault attack, an attacker easily injects the fault at a random position of the state permanently. Hu et al. claimed that a hard fault attack is possible on Trivium [34]. They also claimed that if an attacker prepares the value of some random position of 288 bit states to be permanently 0, then the locations of the injected faulty position attacker can recover the key with a probability of ≮0.2291.

9.5 Research Issues

Crowd computing is an opportunistic system for both machine and crowd. The main goal of our survey is to provide researchers with a better knowledge of the underlying difficulties and solutions for the security and privacy of resource-constrained devices. In the last few years, many researchers have recommended lightweight stream ciphers for practical use in the context of communication, cloud, IoT, and embedded systems. Interestingly, only a few lightweight ciphers, like Sprout, Fruit, and Lizard, have been proposed, and their authors present a technique to reduce the internal state that is provably secure beyond the common attacks. More importantly, the cipher can be trusted by how many cryptanalytic attacks it successfully resists. Here, a few research issues have been identified.

- A complete and integrated security solution is required for crowd computing. Few security solutions are currently available to both end users and service providers, and still security algorithms are facing computational and communication overhead.
- How to balance the contradictory security properties of the lightweight cipher for good service is still being researched and has yet to be solved.
- Constrained devices and social networks suffer many privacy hurdles, like trust, identity privacy, location privacy, and malicious attacks [35]. Thus, privacy is an important concern in resource-constrained devices and social communication so that the illegal usage of data might be resisted.

Lightweight cryptographic primitives are a new approach applicable to securing communication and supporting the confidentiality of important services. The lightweight cryptographic algorithm can also be optimized in order to improve performance according to various hardware platforms and is expected to provide high throughput. The scalability of the stream cipher can be used for better security and performance by reducing the number of internal states, like the Sprout cipher.

9.6 Conclusion

Crowd computing collectively makes the measurements of crowdsourcing, automation, and machine learning. First, we provide a background overview of crowd computing. Here, we show that stream ciphers are suitable for low-end device security. We discuss the taxonomy, characteristics, advantages, and cryptographic features of the lightweight stream cipher. Next, we present the common security components that are used to judge the cipher nonlinear functions, along with tabulating all the results. Then we present symmetric cipher hardware realization results. After that, we discuss the randomness tests of the keystream produced by the Sprout stream cipher using the Dieharder package. In this chapter, we reviewed the common attacks of stream ciphers. However, this research field is still naive and unexplored to a great extent, and various security and privacy challenges are still under research and have yet to be answered.

References

1. Muhammadi, Jafar, Hamid R. Rabiee, and Abbas Hosseini. "A Unified Statistical Framework for Crowd Labeling." *Knowledge and Information Systems* 45, no. 2 (2014), pp. 271–294.
2. Parshotam, Kalpana. "Crowd Computing." In *Proceedings of the South African Institute for Computer Scientists and Information Technologists Conference on*, SAICSIT '13, 2013. East London, South Africa.
3. Lin, Jie, Wei Yu, Nan Zhang, Xinyu Yang, Hanlin Zhang, and Wei Zhao. "A Survey on Internet of Things: Architecture, Enabling Technologies, Security and Privacy, and Applications." *IEEE Internet of Things Journal* 4, no. 5 (2017), pp. 1125–1142.
4. Manifavas, Charalampos, George Hatzivasilis, Konstantinos Fysarakis, and Yannis Papaefstathiou. "A Survey of Lightweight Stream Ciphers for Embedded Systems." *Security and Communication Networks* 9, no. 10 (2015), pp. 1226–1246.
5. Paul, Goutam, and Subhamoy Maitra. *RC4 Stream Cipher and Its Variants*. Boca Raton, FL: CRC Press, 2012.
6. Banik, Subhadeep. "Some Studies on Selected Stream Cipher, Analysis, Fault Attack & Related Results." PhD diss., Indian Statistical Institute Kolkata, 2015.
7. Chakraborty, Abhishek, Bodhisatwa Mazumdar, and Debdeep Mukhopadhyay. "Fibonacci LFSR vs. Galois LFSR: Which Is More Vulnerable to Power Attacks?" In *Security, Privacy, and Applied Cryptography Engineering*, 2014, pp. 14–27. doi: 10.1007/978-3-319-12060-7_2. Pune, India.
8. Zeng, Guang, Wenbao Han, and Kaicheng He. "High Efficiency Feedback Shift Register: σ-LFSR*." Cryptology ePrint Archive, Report 2007/114, 2007.
9. Raizada, Shashwat. "Analysis and Implementation of HC-128 Stream Cipher." PhD diss., Indian Statistical Institute Kolkata, 2015.
10. Maitra, Subhamoy, and Palash Sarkar. "Cryptographically Significant Boolean Functions with Five Valued Walsh Spectra." *Theoretical Computer Science* 276, no. 1–2 (2002), 133–146.
11. Maitra, Subhamoy. "Autocorrelation Properties of Correlation Immune Boolean Functions." *Lecture Notes in Computer Science* vol. 2247, (2001), pp. 242–253, Chennai, India.
12. Nawaz, Yassir, Guang Gong, and Kishan C. Gupta. "Upper Bounds on Algebraic Immunity of Boolean Power Functions." In *Fast Software Encryption*, 2006, pp. 375–389, Graz, Austria.
13. Canteaut, Anne. "Correlation Attack for Stream Ciphers." In *Encyclopedia of Cryptography and Security*, pp. 103. New York: Springer, 2005.
14. Fúster-Sabater, A., and P. Caballero-Gil. "Concatenated Automata in Cryptanalysis of Stream Ciphers." *Lecture Notes in Computer Science* vol. 4173 (2006), pp. 611–616 Perpignan, France.
15. Armknecht, Frederik, and Vasily Mikhalev. "On Lightweight Stream Ciphers with Shorter Internal States." In *Fast Software Encryption*, 2015, pp. 451–470 Istanbul, Turkey.
16. Ghafari, Vahid A., Honggang Hu, and Ying Chen. "Fruit-v2: Ultra-Lightweight Stream Cipher with Shorter Internal State." Cryptology ePrint Archive, Report 2016/355, 2016.
17. Dubrova, Elena, and Martin Hell. "Espresso: A Stream Cipher for 5G Wireless Communication Systems." *Cryptography and Communications* 9, no. 2 (2015), 273–289.
18. "eSTREAM: The ECRYPT Stream Cipher Project." Available at www.ecrypt.eu.org/stream/.

19. Engels, Daniel, Xinxin Fan, Guang Gong, Honggang Hu, and Eric M. Smith. "Hummingbird: Ultra-Lightweight Cryptography for Resource-Constrained Devices." In *Financial Cryptography and Data Security*, (2010), pp. 3–18, Canary Islands, Spain.

20. Yang, Gangqiang, Xinxin Fan, Mark Aagaard, and Guang Gong. "Design Space Exploration of the Lightweight Stream Cipher WG-8 for FPGAs and ASICs." Presented at Proceedings of the Workshop on Embedded Systems Security—WESS '13, 2013. Quebec, Canada.

21. Watanabe, Dai, Kota Ideguchi, Jun Kitahara, Kenichiro Muto, Hiroki Furuichi, and Toshinobu Kaneko. "Enocoro-80: A Hardware Oriented Stream Cipher." Presented at 2008 Third International Conference on Availability, Reliability and Security, 2008 Barcelona, Spain.

22. Leander, Gregor, Christof Paar, Axel Poschmann, and Kai Schramm. "New Lightweight DES Variants." In *Fast Software Encryption*, (2007), pp. 196–210, Luxembourg.

23. Bogdanov, A., L. R. Knudsen, G. Leander, C. Paar, A. Poschmann, M. J. Robshaw, Y. Seurin, and C. Vikkelsoe. "PRESENT: An Ultra-Lightweight Block Cipher." In *Cryptographic Hardware and Embedded Systems—CHES*, (2007), pp. 450–466, Vienna, Austria.

24. Hong, Deukjo, Jaechul Sung, Seokhie Hong, Jongin Lim, Sangjin Lee, Bon-Seok Koo, Changhoon Lee, et al. "HIGHT: A New Block Cipher Suitable for Low-Resource Device." *Lecture Notes in Computer Science*, vol 4249, (2006), pp. 46–59, Yokohama, Japan.

25. Borghoff, Julia, Anne Canteaut, Tim Güneysu, Elif B. Kavun, Miroslav Knezevic, Lars R. Knudsen, Gregor Leander, et al. "PRINCE—A Low-Latency Block Cipher for Pervasive Computing Applications." In *Advances in Cryptology—ASIACRYPT 2012*, 2012, pp. 208–225, Beijing, China.

26. Yang, Gangqiang, Bo Zhu, Valentin Suder, Mark D. Aagaard, and Guang Gong. "The Simeck Family of Lightweight Block Ciphers." *Lecture Notes in Computer Science*, vol 9293 (2015), pp. 307–329, Saint-Malo, France.

27. Fournel, Nicolas, Marine Minier, and Stéphane Ubéda. "Survey and Benchmark of Stream Ciphers for Wireless Sensor Networks." In *Information Security Theory and Practices: Smart Cards, Mobile and Ubiquitous Computing Systems*, 2007, pp. 202–214, Crete, Greece.

28. Barker, E. B., and J. M. Kelsey. "Recommendation for Random Number Generation Using Deterministic Random Bit Generators (revised)." NIST, 2007. doi: 10.6028/nist.sp.800-90.

29. Brown, Robert G. "Dieharder: A Random Number Test Suite (version 3.31)." 2013. Available at http://webhome.phy.duke.edu/~rgb/General/dieharder.php.

30. Hellman, M. "A Cryptanalytic Time-Memory Trade-Off." *IEEE Transactions on Information Theory* 26, no. 4 (1980), pp. 401–406.

31. Teo, Sui G. "Analysis of Nonlinear Sequences and Stream Ciphers." PhD diss., Queensland University of Technology, 2013.

32. Siegenthaler, T. "Decrypting a Class of Stream Ciphers Using Ciphertext Only." *IEEE Transactions on Computers*, C-34, no. 1 (January 1985), pp. 81–85.

33. Dinur, Itai, and Adi Shamir. "Cube Attacks on Tweakable Black Box Polynomials." In *Advances in Cryptology—EUROCRYPT 2009*, 2009, pp. 278–299, Cologne, Germany.

34. Hu, Yu-pu, Feng-rong Zhang, and Wen-zheng Zhang. "Hard Fault Analysis of Trivium." *Information Sciences* 229 (2013), pp. 142–158.

35. Haus Michael, , Muhammad Waqas, and Aaron Yi Ding. "Security and Privacy in Device-to-Device (D2D) Communication: A Review." *IEEE Communications Surveys & Tutorials* 19, no. 2 (2017), pp. 1054–1079.

Chapter 10

A New Cryptanalysis Method of 4-Bit Crypto S-Boxes in Crowd Computing

Sankhanil Dey and Ranjan Ghosh

Contents

10.1 Introduction

Crowd computing or computing with crowds was introduced to literature in 2009 [1,2]. The crowd has been used in solving human problems with web and mobile computation. The essential component has been the crowd, but one of the most important and exclusive components in crowd computing has been data and voice transfer since human resources depend on a safe and secure data transfer through the web and Android [3,4]. Secure data transfer needs a good encryption and decryption algorithm or encryption standard in cryptology. An encryption and decryption algorithm must contain a substitution box (S-box) for nonlinear substitution of plaintext bits and nonlinear substitution of S-boxes [5,6]. In this chapter, an existing analysis algorithm of 4-bit S-boxes has been reviewed in brief and a new analysis technique to analyze 4-bit S-boxes has been introduced in detail since data security is now a prime concern in crowd computing.

The exclusive-or or XOR operation is defined to be a linear operation in cryptography. Linear operations are used to give two exact values, 0 and 1, in operation between two same and different bits, respectively, in Boolean logic or switching logic. So if a linear relation exists between all 4-bit plaintext bit patterns and the corresponding 4-bit ciphertext bit patterns, then the existing relation between them is easy to determine. The idea of using linear relations to analyze the randomization property of a cipher was introduced by Matsui in 1994 for cryptanalysis of the reduced-round Data Encryption Standard (DES) cipher [7] and DES algorithm [8]. Later, Heys [9] extended the idea toward 4-bit S-boxes in his tutorial on linear and differential cryptanalysis of 4-bit S-boxes and his contribution to substitution-permutation networks (SPNs) [10].

A 4-bit S-box consists of 16 array elements whose indices are considered 4-bit inputs corresponding to sequential hex values from 0 to f. The output data corresponding to each array index are supposed to have 4-bit sequential or nonsequential hex values between 0 and f. Such an S-box with 4-bit input and 4-bit output is called a bijective S-box [11]. Nonbijective 4-bit S-boxes are those whose inputs may consist of a number of bits more than 4. For all 4-bit S-boxes, the four input vectors (IPVs) are the same and the output would be composed of four Boolean functions (BFs) giving four 16-bit output column vectors whose row-wise 4 bits assume hex values lying between 0 and *f*. The number of possible S-boxes is obtained as factorial 16 (16!) following the permutation of 16 hex digits between 0 and *f*. A 4-bit S-box can also be represented by a four-valued Boolean function following the norms of presentation of multivalued BFs [5].

In linear cryptanalysis of 4-bit crypto S-boxes, every 4-bit linear relation has been tested for a particular 4-bit crypto S-box. The presence of each 4-bit unique linear relation is checked by the satisfaction of each of them for all sixteen 4-bit unique input bit patterns and corresponding 4-bit output bit patterns, generated from the index of each element and each element, respectively, of that particular crypto S-box. If they are satisfied 8 times out of 16 operations for all 4-bit unique input bit patterns and corresponding 4-bit output bit patterns, then the existence of the 4-bit linear equation is at a stake. The probability of both the presence and absence of a 4-bit linear relation is (= 8/16) ½. If a 4-bit linear equation is satisfied 0 times, then it can be concluded that the given 4-bit linear relation is absent for that particular 4-bit crypto S-box. If a 4-bit linear equation is satisfied 16 times, then it can also be concluded that the given 4-bit linear relation is present for that particular 4-bit crypto S-box. In both cases, full information is adverted to the cryptanalysts. The concept of probability bias was introduced to predict the randomization ability of that 4-bit S-box from the probability of the presence or absence of unique 4-bit linear relations. The result is better for cryptanalysts if the probability of the presence or absence of unique 4-bit linear equations is far from ½ or near to 0 or 1. If the probabilities of the presence or absence of all unique 4-bit linear relations are ½ or close to ½, then the 4-bit crypto S-box is said to be linear cryptanalysis immune, since the existence of maximum 4-bit linear relations for that 4-bit crypto S-box is hard to predict [9,10]. Heys also introduced the concept of the linear approximation table (LAT), in which the number of times each 4-bit unique linear relation has been satisfied for all 16 unique 4-bit input bit patterns and corresponding 4-bit output bit patterns of a crypto S-box has been noted. The result is better for a cryptanalyst if the number of 8s in the table is less. If numbers of 8s is much more than the other numbers in the table, then the 4-bit crypto S-box has been said to be more linear cryptanalysis immune [9,10].

In another look, an input S-box can be decomposed into four 4-bit IPVs with decimal equivalents 255 for the fourth IPV, 3,855 for the third IPV, 13,107 for the second IPV, and 21,845 for the first IPV, respectively. The S-box can also be decomposed into four 4-bit output Boolean functions (OPBFs). Each IPV can be denoted as an input variable of a linear relation, OPBF as an output variable (OPV), and + as an XOR operation. Linear relations have been checked for satisfaction, and 16-bit OPVs due to linear relations have been checked for balancedness. Balanced OPVs indicate that out of 16 bits of IPVs and OPBFs, 8 bits satisfy the linear relation and 8 bits are out of satisfaction, that is, best uncertainty. A total of 256 4-bit linear relations have been operated on four 16-bit IPVs and four 16-bit OPBFs, and 256 OPVs have been generated. The count of the number of 1s in OPVs has been put in the LAT. The better the number of 8s in the LAT, the better the S-box security [9,10]. The concept is reviewed in brief in Section 10.4.

In this chapter, a new technique to find the existing linear relations or linear approximations for a particular 4-bit S-box is reviewed in detail. If the nonlinear part (NP) of the algebraic normal form (ANF) equation of a 4-bit OPBF is absent or calculated to be 0, then the equation is termed a linear relation or approximation. Searching for a number of existing linear relations through this method, we ended up with a number of existing linear relations; that is, the goal to conclude the security of a 4-bit crypto S-box was attended to in a very lucid manner by this method. The method is reviewed in Section 10.5.

A brief literature study is given in Section 10.2. The definition of crowd computing is described in Section 10.3. An ANF of 4-bit BFs and linear cryptanalysis of 4-bit S-boxes are introduced in a very lucid manner in Section 10.4. The new cryptanalysis method or linear approximation analysis (LAA) is described in Section 10.5. The pseudocode of a new algorithm is also given in Section 10.5. An analysis of the results is given in Section 10.6. Section 10.7 concludes the chapter, with an analysis of thirty-two 4-bit crypto S-boxes of the DES given in the appendix.

10.2 Literature Survey

In this section, an exhaustive relevant literature survey with specific references is introduced to crypto-literature, including cryptography and cryptology (Section 10.2.1) and linear cryptanalysis (Section 10.2.2).

10.2.1 Cryptography and Cryptology

At the end of the twentieth century, a bible of cryptography was introduced [12]. The various concepts involved in cryptography, as well as some information on cryptanalysis, was provided to the crypto-community in the late nineties [13]. A simplified version of DES that has the architecture of DES but much fewer rounds and bits was also proposed at the same time. The cipher has also been improved for educational purposes [14]. Later, in the early twenty-first century, an organized pathway toward learning how to cryptanalyze was charted [15]. Almost at the same time, a new cipher as a candidate for the new Advanced Encryption Standard (AES), with the main concepts and issues involving block cipher design and cryptanalysis, was also proposed [16] as a measure of cipher strength. A vital preliminary introduction to cryptanalysis has also been introduced to cryptanalysts [17]. At the same time, a somewhat similar notion [17], but using a more descriptive approach and focused on linear cryptanalysis and differential cryptanalysis of a given SPN cipher, was elaborated [18]. Particularly, it discussed DES-like ciphers that had been extended with it [19]. Comparison of modes of operations, such as CBC, CFB, OFB, and ECB, were also elaborated [20]. A new cipher, called "Camelia," was introduced with its cryptanalysis technique to demonstrate the strength of the cipher [21]. The history of commercial computer cryptography and classical ciphers and the effect of cryptography on society were also introduced in this queue [22]. The requirements of a good cryptosystem and cryptanalysis were demonstrated later [23]. A description of the new AES by Rijndael, providing good insight into many creative cryptographic techniques that increase cipher strength, was included in the literature. A bit later, a highly mathematical path to explain cryptologic concepts was introduced [24]. Investigation of the security of Ron Rivest's DESX construction, a cheaper alternative to triple DES, was elaborated [25]. A nice provision to an encyclopedic look at the design, the analysis and applications of cryptographic techniques were later depicted [11], and last but not least, a good explanation on why cryptography has been hard, as well as the issues that cryptographers have to consider in designing ciphers, was elaborated [26]. The Simplified Data Encryption Standard (S-DES) is an educational algorithm similar to the DES but with much smaller parameters [27,28]. The technique to analyze S-DES using linear cryptanalysis and differential cryptanalysis has been of interest in the crypto-community [27,28]. The encryption and decryption algorithm or cipher of the two-fish algorithm was introduced to the crypto-community, and a cryptanalysis of the said cipher was elaborated to be a part of advanced encryption algorithm proposals [29].

10.2.2 Some Old and Recent References on Linear Cryptanalysis

The cryptanalysis technique for 4-bit crypto S-boxes using linear relations among four 4-bit IPVs and four output 4-bit Boolean functions (OPBFs) of a 4-bit S-box has been termed linear cryptanalysis of 4-bit crypto S-boxes [9,10]. Another technique to analyze the security of a 4-bit crypto S-box using all the possible differences is called differential cryptanalysis of 4-bit crypto S-boxes [9,10].The search for the best characteristic in linear cryptanalysis and the maximal weight path in a directed graph, and correspondence between them, is elaborated with an example [30]. It has been proposed that a correlation matrix be used as a natural representation to understand and describe the mechanism of linear cryptanalysis [31]. The method described in [7] was also formalized and showed that at the structural level, linear cryptanalysis is very similar to differential cryptanalysis. It was also used for further exploration into linear cryptanalysis [32]. It has been provided with a generalization of linear cryptanalysis and suggests that IDEA and SAFER K-64 are secure against such generalizations [33]. A survey had been made to show the the use of multiple linear approximations in cryptanalysis to improve efficiency and reduce the amount of data required for cryptanalysis in certain circumstances [34]. Cryptanalysis of the DES cipher with linear relations [7] and an improved version of the said cryptanalysis [7] with 12 computers was reported later [8]. An implementation of Matsui's linear cryptanalysis of DES with strong emphasis on efficiency has also been reported [35]. In the early days of this century, a cryptanalytic attack based on multiple linear approximations of the AES candidate "Serpent" was also described [36]. Later, a technique to prove security bounds against linear and differential cryptanalytic attacks using mixed-integer linear programming (MILP) was elaborated [37]. After this, on the strength of two variants of reduced-round lightweight block cipher, SIMON-32 and SIMON-48 were tested against linear cryptanalysis and presented the optimum possible results [38]. Almost at the same time, the strength of another lightweight block cipher, SIMECK, was tested against linear cryptanalysis [39]. The fault analysis of the lightweight block cipher SPECK and linear cryptanalysis with zero statistical correlation between plaintext and the respective ciphertext of the reduced-round lightweight block cipher SIMON were also recently introduced to test the cipher's strength against cryptanalytic attacks [39–41].

10.3 Brief Review of Crowd Computing

The term *crowd computing* is a new era of computation that means computation with a crowd. If a huge and tedious digitization work has been done with the help of the contribution of a large crowd of people through web or Android, then the computation work is called crowd computing. Let us look at an example of cloud computing with human sources. If many remote or human resources conduct a huge computation with small contributions with the help of the cloud, then it is also been called crowd computing. This era began with crowd sourcing in 2009 [14] and has been elaborated in various ways, including the distribution of human intelligence tasks to mobile devices, cloud computing with humans, human problem solving with large numbers of people using computers, and broadly, as a set of human interaction tools for idea exchange and nonhierarchical decision making. From the literature, four common characteristics can be identified to define the boundaries of the term: participation by a crowd of humans, interaction with computing technology, activity that is predetermined by the initiator or application itself, and the execution of tasks by the crowd utilizing innate human capabilities. Figure 10.1 shows a method of crowd computing.

Recently, adversaries have played a heinous role in cyber-attacks. Data transfer or the secure transfer of computed data through the Internet was an insecure task before the introduction of cryptography. There are many examples in the past of cyber-attacks. Because of this, the Lloyds Bank proposed IBM for an encryption algorithm for ATM transactions in the late sixties. The Lucifer algorithm came

Figure 10.1 Crowd computing with human sources.

from this negotiation. Later, NIST in the United States called for a government encryption standard. DES came out of that. It proved to be backdated and insecure by Matsui by linear cryptanalysis.

In this chapter, a new algorithm for the cryptanalysis of substitution boxes is proposed with its minimum time complexity. All encryption and decryption algorithms and their substitution boxes have been tested for better security, and this will make the data transfer of crowd computing a more secure job. Although crowd computing was introduced in 2009, secure data transfer is a major part of crowd computing with the cloud and fog computing. The reliability of software also depends on secure data transmission through the Internet. So, cryptography has a role in crowd computing as well as in cloud and fog computing. Secure data transfer needs a secure encryption algorithm as well as a secure substitution box. The security of substitution boxes has been tested through cryptanalysis. So, linear cryptanalysis and LAA have been reviewed and introduced and respectively proved to be of utmost importance in crowd computing.

10.4 Review of Linear Cryptanalysis of 4-Bit Crypto S-Boxes [9,10]

The given 4-bit crypto S-box is described in Section 10.4.1. The relation between 4-bit S-boxes and 4-bit BFs and linear approximations is described Sections 10.4.2 and 10.4.3, respectively. LAT is illustrated in Section 10.4.4. An algorithm for linear cryptanalysis with time complexity analysis is described in Section 10.4.5.

10.4.1 4-Bit Crypto S-Boxes

A 4-bit crypto S-box can be written shown in Table 10.1, where each element of the first row of Table 10.1, titled "Index," is the position of each element of the S-box within the given S-box, and the elements of the second row, titled "S-box," are the

Table 10.1   4-Bit Crypto S-Box

Row	Column	1	2	3	4	5	6	7	8	9	A	B	C	D	E	F	G
1	Index	0	1	2	3	4	5	6	7	8	9	A	B	C	D	E	F
2	S-box	E	4	D	1	2	F	B	8	3	A	6	C	5	9	0	7

elements of the given substitution box. It can be concluded that the first row is fixed for all possible crypto S-boxes. The values of each element of the first row are distinct, unique, and vary between 0 and F in hex. The values of each element of the second row of a crypto S-box are also distinct and unique and also vary between 0 and F in hex. The values of the elements of the fixed first row are sequential and monotonically increasing, whereas for the second row they can be sequential, partly sequential, or nonsequential. Here the given substitution box is the first 4-bit S-box of the first S-box out of eight in the DES [5,6].

10.4.2 Relation between 4-Bit S-Boxes and 4-Bit Boolean Functions

The index of each element of a 4-bit crypto S-box and the element itself are hexadecimal numbers that can be converted into a 4-bit sequence, given in columns 1 through G of rows 1 and 6 under the rows titled "Index" and "S-box," respectively. Rows 2 through 5 and rows 7 through A of each column from 1 through G of Table 10.2 show the 4-bit sequences of the corresponding hexadecimal numbers of the index of each element of the given crypto S-box and each element of the crypto S-box itself. Each row from 2 through 5 and 7 through A from columns 1 through G constitutes a 16-bit sequence that is 16-bit-long IPVs and 4-bit OPBFs, respectively. Columns 1 through G of row 2 are the fourth IPV, row 3 is the third IPV, row 4 is the second IPV, and row 5 is the first IPV, whereas columns 1 through G of row 7 are the fourth OPBF, row 8 is the third OPBF, row 9 is the second OPBF, and row A is the first OPBF [5]. The decimal equivalent of each IPV and OPBF is noted in column H of the respective rows.

10.4.3 4-Bit Linear Relations

The elements of the input S-box are shown under the column labeled "I," and the IPVs are shown under the "IPVs" field and subsequently under columns 1, 2, 3, and 4. The fourth IPV is depicted under column 4, the third IPV is under column 3, the second IPV is under column 2, and the first IPV is under column 1. The elements of the S-box are shown under the "SB" column, and the output 4-bit BFs are shown under the "OPBFs" field and subsequently under columns 1, 2, 3 and 4. The fourth OPBF is depicted under column 4, the third OPBF is under column 3, the second OPBF is under column 2, and the first OPBF is under column 1 of Table 10.3.

Table 10.2 Decomposition of 4-Bit Input S-Box and the Given S-Box (First 4-Bit S-Box of First S-Box Out of Eight for DES) into 4-Bit BFs

Row	Column	1	2	3	4	5	6	7	8	9	A	B	C	D	E	F	G	H: Decimal Equivalent
1	Index	0	1	2	3	4	5	6	7	8	9	A	B	C	D	E	F	
2	IPV4	0	0	0	0	0	0	0	0	1	1	1	1	1	1	1	1	00255
3	IPV3	0	0	0	0	1	1	1	1	0	0	0	0	1	1	1	1	03855
4	IPV2	0	0	1	1	0	0	1	1	0	0	1	1	0	0	1	1	13107
5	IPV1	0	1	0	1	0	1	0	1	0	1	0	1	0	1	0	1	21845
6	S-box	E	4	D	1	2	F	B	8	3	A	6	C	5	9	0	7	
7	OPBF4	1	0	1	0	0	1	1	1	0	1	0	1	0	1	0	0	42836
8	OPBF3	1	1	1	0	0	1	0	0	0	0	1	1	1	0	0	1	58425
9	OPBF2	1	0	0	0	1	1	1	0	1	1	1	0	0	0	0	1	36577
A	OPBF1	0	0	1	1	0	1	1	0	1	0	0	0	1	1	0	1	13965

Table 10.3   IPVs and OPBFs for Given S-Box

I	IPVs				S B	OPBFs			
	4	3	2	1		4	3	2	1
0	0	0	0	0	E	1	1	1	0
1	0	0	0	1	4	0	1	0	0
2	0	0	1	0	D	1	1	0	1
3	0	0	1	1	1	0	0	0	1
4	0	1	0	0	5	0	1	0	1
5	0	1	0	1	9	1	0	0	1
6	0	1	1	0	0	0	0	0	0
7	0	1	1	1	7	0	1	1	1
8	1	0	0	0	2	0	0	1	0
9	1	0	0	1	F	1	1	1	1
A	1	0	1	0	B	1	0	1	1
B	1	0	1	1	8	1	0	0	0
C	1	1	0	0	3	0	0	1	1
D	1	1	0	1	A	1	0	1	0
E	1	1	1	0	6	0	1	1	0
F	1	1	1	1	C	1	1	0	0

The input equations (IPEs) are all possible XORed terms that can be formed using four IPVs, 4, 3, 2, and 1. On the other hand, output equations (OPEs) are possible XORed terms that can be formed using four OPVs, 4, 3, 2, and 1. All possible IPEs and OPEs are listed under the column and also row (IPE = OPE) from rows 2 through H and columns 1 through G, respectively. Each cell is a linear equation equating IPE to OPE. For example, $L_{1+2+4,2+3}$ is the linear equation formed by IPE 1 + 2 + 3 (i.e., the XORed combination of three IPVs, 1, 2, and 4) and OPE 2 + 3 (i.e., the XORed combination of two OPBFs, 2 and 3). The 256 possible 4-bit linear equations are shown in Table 10.4.

10.4.4 Linear Approximation Table [6]

According to Heys, each linear equation is tested for each of sixteen 4-bit patterns shown in each row under the "IPVs" field and subsequently under columns 1, 2, 3,

Table 10.4 Two Hundred Fifty-Six 4-Bit Linear Equations with Input Equations (IPE) and Output Equations (OPE)

Row	Column	1	2	3	4	5	6	7	8	9
		0	1	2	3	4	1 + 2	1 + 3	1 + 4	2 + 3
1	IPE = OPE	0	1	2	3	4	1 + 2	1 + 3	1 + 4	2 + 3
2	0	$L_{0,0}$	$L_{0,1}$	$L_{0,2}$	$L_{0,3}$	$L_{0,4}$	$L_{0,1+2}$	$L_{0,1+3}$	$L_{0,1+4}$	$L_{0,2+3}$
3	1	$L_{1,0}$	$L_{1,1}$	$L_{1,2}$	$L_{1,3}$	$L_{1,4}$	$L_{1,1+2}$	$L_{1,1+3}$	$L_{1,1+4}$	$L_{1,2+3}$
4	2	$L_{2,0}$	$L_{2,1}$	$L_{2,2}$	$L_{2,3}$	$L_{2,4}$	$L_{2,1+2}$	$L_{2,1+3}$	$L_{2,1+4}$	$L_{2,2+3}$
5	3	$L_{3,0}$	$L_{3,1}$	$L_{3,2}$	$L_{3,3}$	$L_{3,4}$	$L_{3,1+2}$	$L_{3,1+3}$	$L_{3,1+4}$	$L_{3,2+3}$
6	4	$L_{4,0}$	$L_{4,1}$	$L_{4,2}$	$L_{4,3}$	$L_{4,4}$	$L_{4,1+2}$	$L_{4,1+3}$	$L_{4,1+4}$	$L_{4,2+3}$
7	1 + 2	$L_{1+2,0}$	$L_{1+2,1}$	$L_{1+2,2}$	$L_{1+2,3}$	$L_{1+2,4}$	$L_{1+2,1+2}$	$L_{1+2,1+3}$	$L_{1+2,1+4}$	$L_{1+2,2+3}$
8	1 + 3	$L_{1+3,0}$	$L_{1+3,1}$	$L_{1+3,2}$	$L_{1+3,3}$	$L_{1+3,4}$	$L_{1+3,1+2}$	$L_{1+3,1+3}$	$L_{1+3,1+4}$	$L_{1+3,2+3}$
9	1 + 4	$L_{1+4,0}$	$L_{1+4,1}$	$L_{1+4,2}$	$L_{1+4,3}$	$L_{1+4,4}$	$L_{1+4,1+2}$	$L_{1+4,1+3}$	$L_{1+4,1+4}$	$L_{1+4,2+3}$
A	2 + 3	$L_{2+3,0}$	$L_{2+3,1}$	$L_{2+3,2}$	$L_{2+3,3}$	$L_{2+3,4}$	$L_{2+3,1+2}$	$L_{2+3,1+3}$	$L_{2+3,1+4}$	$L_{2+3,2+3}$
B	2 + 4	$L_{2+4,0}$	$L_{2+4,1}$	$L_{2+4,2}$	$L_{2+4,3}$	$L_{2+4,4}$	$L_{2+4,1+2}$	$L_{2+4,1+3}$	$L_{2+4,1+4}$	$L_{2+4,2+3}$
C	3 + 4	$L_{3+4,0}$	$L_{3+4,1}$	$L_{3+4,2}$	$L_{3+4,3}$	$L_{3+4,4}$	$L_{3+4,1+2}$	$L_{3+4,1+3}$	$L_{3+4,1+4}$	$L_{3+4,2+3}$
D	1 + 2 + 3	$L_{1+2+3,0}$	$L_{1+2+3,1}$	$L_{1+2+3,2}$	$L_{1+2+3,3}$	$L_{1+2+3,4}$	$L_{1+2+3,1+2}$	$L_{1+2+3,1+3}$	$L_{1+2+3,1+4}$	$L_{1+2+3,2+3}$
E	1 + 2 + 4	$L_{1+2+4,0}$	$L_{1+2+4,1}$	$L_{1+2+4,2}$	$L_{1+2+4,3}$	$L_{1+2+4,4}$	$L_{1+2+4,1+2}$	$L_{1+2+4,1+3}$	$L_{1+2+4,1+4}$	$L_{1+2+4,2+3}$
F	1 + 3 + 4	$L_{1+3+4,0}$	$L_{1+3+4,1}$	$L_{1+3+4,2}$	$L_{1+3+4,3}$	$L_{1+3+4,4}$	$L_{1+3+4,1+2}$	$L_{1+3+4,1+3}$	$L_{1+3+4,1+4}$	$L_{1+3+4,2+3}$
G	2 + 3 + 4	$L_{2+3+4,0}$	$L_{2+3+4,1}$	$L_{2+3+4,2}$	$L_{2+3+4,3}$	$L_{2+3+4,4}$	$L_{2+3+4,1+2}$	$L_{2+3+4,1+3}$	$L_{2+3+4,1+4}$	$L_{2+3+4,2+3}$
H	1 + 2 + 3 + 4	$L_{1+2+3+4,0}$	$L_{1+2+3+4,1}$	$L_{1+2+3+4,2}$	$L_{1+2+3+4,3}$	$L_{1+2+3+4,4}$	$L_{1+2+3+4,1+2}$	$L_{1+2+3+4,1+3}$	$L_{1+2+3+4,1+4}$	$L_{1+2+3+4,2+3}$

(Continued)

Table 10.4 (Continued)   Two Hundred Fifty-Six 4-Bit Linear Equations with Input Equations (IPE) and Output Equations (OPE)

Row	Column	A	B	C	D	E	F	G
2	0	$L_{0,2+4}$	$L_{0,3+4}$	$L_{0,1+2+3}$	$L_{0,1+2+4}$	$L_{0,1+3+4}$	$L_{0,2+3+4}$	$L_{0,1+2+3+4}$
3	1	$L_{1,2+4}$	$L_{1,3+4}$	$L_{1,1+2+3}$	$L_{1,1+2+4}$	$L_{1,1+3+4}$	$L_{1,2+3+4}$	$L_{1,1+2+3+4}$
4	2	$L_{2,2+4}$	$L_{2,3+4}$	$L_{2,1+2+3}$	$L_{2,1+2+4}$	$L_{2,1+3+4}$	$L_{2,2+3+4}$	$L_{2,1+2+3+4}$
5	3	$L_{3,2+4}$	$L_{3,3+4}$	$L_{3,1+2+3}$	$L_{3,1+2+4}$	$L_{3,1+3+4}$	$L_{3,2+3+4}$	$L_{3,1+2+3+4}$
6	4	$L_{4,2+4}$	$L_{4,3+4}$	$L_{4,1+2+3}$	$L_{4,1+2+4}$	$L_{4,1+3+4}$	$L_{4,2+3+4}$	$L_{4,1+2+3+4}$
7	1 + 2	$L_{1+2,2+4}$	$L_{1+2,3+4}$	$L_{1+2,1+2+3}$	$L_{1+2,1+2+4}$	$L_{1+2,1+3+4}$	$L_{1+2,2+3+4}$	$L_{1+2,1+2+3+4}$
8	1 + 3	$L_{1+3,2+4}$	$L_{1+3,3+4}$	$L_{1+3,1+2+3}$	$L_{1+3,1+2+4}$	$L_{1+3,1+3+4}$	$L_{1+3,2+3+4}$	$L_{1+3,1+2+3+4}$
9	1 + 4	$L_{1+4,2+4}$	$L_{1+4,3+4}$	$L_{1+4,1+2+3}$	$L_{1+4,1+2+4}$	$L_{1+4,1+3+4}$	$L_{1+4,2+3+4}$	$L_{1+4,1+2+3+4}$
A	2 + 3	$L_{2+3,2+4}$	$L_{2+3,3+4}$	$L_{2+3,1+2+3}$	$L_{2+3,1+2+4}$	$L_{2+3,1+3+4}$	$L_{2+3,2+3+4}$	$L_{2+3,1+2+3+4}$
B	2 + 4	$L_{2+4,2+4}$	$L_{2+4,3+4}$	$L_{2+4,1+2+3}$	$L_{2+4,1+2+4}$	$L_{2+4,1+3+4}$	$L_{2+4,2+3+4}$	$L_{2+4,1+2+3+4}$
C	3 + 4	$L_{3+4,2+4}$	$L_{3+4,3+4}$	$L_{3+4,1+2+3}$	$L_{3+4,1+2+4}$	$L_{3+4,1+3+4}$	$L_{3+4,2+3+4}$	$L_{3+4,1+2+3+4}$
D	1 + 2 + 3	$L_{1+2+3,2+4}$	$L_{1+2+3,3+4}$	$L_{1+2+3,1+2+3}$	$L_{1+2+3,1+2+4}$	$L_{1+2+3,1+3+4}$	$L_{1+2+3,2+3+4}$	$L_{1+2+3,1+2+3+4}$
E	1 + 2 + 4	$L_{1+2+4,2+4}$	$L_{1+2+4,3+4}$	$L_{1+2+4,1+2+3}$	$L_{1+2+4,1+2+4}$	$L_{1+2+4,1+3+4}$	$L_{1+2+4,2+3+4}$	$L_{1+2+4,1+2+3+4}$
F	1 + 3 + 4	$L_{1+3+4,2+4}$	$L_{1+3+4,3+4}$	$L_{1+3+4,1+2+3}$	$L_{1+3+4,1+2+4}$	$L_{1+3+4,1+3+4}$	$L_{1+3+4,2+3+4}$	$L_{1+3+4,1+2+3+4}$
G	2 + 3 + 4	$L_{2+3+4,2+4}$	$L_{2+3+4,3+4}$	$L_{2+3+4,1+2+3}$	$L_{2+3+4,1+2+4}$	$L_{2+3+4,1+3+4}$	$L_{2+3+4,2+3+4}$	$L_{2+3+4,1+2+3+4}$
H	1 + 2 + 3 + 4	$L_{1+2+3+4,2+4}$	$L_{1+2+3+4,3+4}$	$L_{1+2+3+4,1+2+3}$	$L_{1+2+3+4,1+2+4}$	$L_{1+2+3+4,1+3+4}$	$L_{1+2+3+4,2+3+4}$	$L_{1+2+3+4,1+2+3+4}$

and 4, and the corresponding sixteen 4-bit patterns under the "OPBFs" field and subsequently under columns 1, 2, 3, and 4. If a linear equation is satisfied 8 times out of 16, then the existence of the linear equation is highly unpredictable. That is, the probability is ½. If the number of satisfaction of each linear equation is noted in the respective cells of Table 10.3, then it is called LAT. The LAT for the given S-box is shown in Table 10.5.

10.4.5 Pseudocode of Algorithm with Time Complexity Analysis of Linear Cryptanalysis of 4-Bit Crypto S-Boxes

The algorithm to execute the linear cryptanalysis for 4-bit crypto S-boxes following Heys [9,10] considers 4-bit Boolean variables Ai and Bj, where i and j are the decimal indices varying from 0 to 15 and Ai and Bj are take the corresponding bit values from [0000] to [1111]. The algorithm to fill the (16×16) elements of the LAT is

```
for(i=0;i<16;i++){
    A=0;
    for(k=0;k<16;k++) A=A+(Ai0.Xk0+Ai1.Xk1+Ai2.Xk2+Ai3.Xk3)%2;
    for(j=0;j<16;j++){
        B=0;
        for(k=0;k<16;k++)B= B+(Bj0.Yk0+Bj1.Yk1+Bj2.Yk2+Bj3.Yk3)%2;
        Sij = (A+B)%2;
        if (Sij==0) Cij++; Nij = Cij - 8;
    }
}
```

Time complexity of the given algorithm. Since the pseudocode contains two nested loops, the time complexity of the given algorithm is $O(n^2)$.

10.5 Linear Approximation Analysis

A crypto 4-bit S-box (first 4-bit S-box out of thirty-two 4-bit S-boxes for DES) is described in Section 10.5.1. The table for four IPVs, 4-bit OPBFs, and the corresponding ANFs is given in Section 10.5.2. The analysis is described in Section 10.5.3, and the result of analysis is given Section 10.5.4.

10.5.1 4-Bit Crypto S-Boxes

A 4-bit crypto S-box can be written shown in Table 10.6, where each element of the first row of Table 10.6, titled "Index," is the position of each element of the S-box within the given S-box, and the elements of the second row, titled "S-box," are the elements of the given substitution box. It can be concluded that the first

Table 10.5 Linear Approximation Table for the Given S-Box

Input Sum	Output Sum															
	0	1	2	3	4	5	6	7	8	9	A	B	C	D	E	F
0	+8	0	0	0	0	0	0	0	0	0	0	0	0	0	0	0
1	0	0	-2	-2	0	0	-2	+6	+2	+2	0	0	+2	+2	0	0
2	0	0	-2	-2	0	0	-2	-2	0	0	+2	+2	0	0	-6	+2
3	0	0	0	0	0	0	0	0	+2	-6	-2	-2	+2	+2	-2	-2
4	0	+2	0	-2	-2	-4	-2	0	0	-2	0	+2	+2	-4	+2	0
5	0	-2	-2	0	-2	0	+4	+2	-2	0	-4	+2	0	-2	-2	0
6	0	+2	-2	+4	+2	0	0	+2	0	-2	+2	+4	-2	0	0	-2
7	0	-2	0	+2	+2	-4	+2	0	-2	0	+2	0	+4	+2	0	+2
8	0	0	0	0	0	0	0	0	-2	+2	+2	-2	+2	-2	-2	-6
9	0	0	-2	-2	0	0	-2	-2	-4	0	-2	+2	0	+4	+2	-2
A	0	+4	-2	+2	-4	0	+2	0	+2	+2	0	0	+2	+2	0	0
B	0	+4	0	-4	+4	0	+4	0	0	0	0	0	0	0	0	0
C	0	-2	+4	-2	-2	0	+2	0	+2	0	+2	44	+2	+2	0	-2
D	0	+2	+2	0	-2	0	0	0	-4	-2	+2	0	0	0	0	+2
E	0	+2	+2	0	-2	+4	0	+2	-2	0	0	-2	-4	+2	-2	0
F	0	-2	-4	-2	-2	0	+2	0	0	-2	+4	-2	-2	0	+2	0

Table 10.6   4-Bit Crypto S-Box

Row	Column	1	2	3	4	5	6	7	8	9	A	B	C	D	E	F	G
1	**Index**	0	1	2	3	4	5	6	7	8	9	A	B	C	D	E	F
2	**S-box**	E	4	D	1	2	F	B	8	3	A	6	C	5	9	0	7

row is fixed for all possible crypto S-boxes. The values of each element of the first row are distinct, unique, and vary between 0 and F in hex. The values of each element of the second row of a crypto S-box are also distinct and unique and also vary between 0 and F in hex. The values of the elements of the fixed first row are sequential and monotonically increasing, whereas for the second row they can be sequential, partly sequential, or nonsequential. Here the given substitution box is the first 4-bit S-box of the first S-box out of eight for the DES [5,6].

10.5.2 Input Vectors, Output Boolean Functions, and Algebraic Normal Forms

The elements of the input S-box are shown under the "ISB" column, and the IPVs are under the "IPV" field and subsequently under columns 1, 2, 3, and 4. The fourth IPV is depicted under column 4, the third IPV is under column 3, the second IPV is under column 2, and the first IPV is under column 1. The elements of the S-box are shown under the "OSB" column, and the output 4-bit BFs are shown under the "OPBF" field and subsequently under columns 1, 2, 3, and 4. The fourth OPBF is depicted under column 4, the third OPBF is under column 3, the second OPBF is under column 2, and the first OPBF is under column 1. The corresponding ANFs for four OPBFs, OPBF-4th, OPBF-3rd, OPBF-2nd, and OPBF-1st, are under the "ANF" field and subsequently under columns 4, 3, 2, and 1, respectively, of Table 10.7.

10.5.3 Linear Approximation Analysis

An ANF equation is a linear equation or linear approximation if the NP (i.e., the XORed value of all product terms of 4-bit ANF equation for the corresponding 4-bit values of IPVs, with columns 4, 3, 2, and 1) is 0 and the linear part (LP) for the corresponding 4-bit values of IPVs, with columns 4, 3, 2, and 1, is equal to the corresponding BF bit values. The corresponding ANF coefficients of OPBFs F(4), F(3), F(2), and F(1) are given under rows ANF(F4), ANF(F3), ANF(F2), and ANF(F1), respectively, from rows 2 through 5 and columns 4 through J. Column 4 of rows 2 through 5 gives the value of the constant coefficient (a_0 according to Equation 2) of ANF(F4), ANF(F3), ANF(F2), and ANF(F1), respectively. Columns 5 through 8 of rows 2 through 5 give the value of the respective linear coefficients more specifically, a_1, a_2, a_3, and a_4 (according to Equation 2) of ANF(F4), ANF(F3), ANF(F2),

Table 10.7 Input and Output Boolean Functions with Corresponding ANF Coefficients of the Given S-Box

ISB	IPV 4321	OSB	OPBF 4321	ANF 4321
0	0000	E	1110	1110
1	0001	4	0100	1010
2	0010	D	1101	0011
3	0011	1	0001	1100
4	0100	2	0010	1101
5	0101	F	1111	0110
6	0110	B	1011	0111
7	0111	8	1000	0011
8	1000	3	0011	1010
9	1001	A	1010	0110
A	1010	6	0110	1010
B	1011	C	1100	1000
C	1100	5	0101	0101
D	1101	9	1001	0010
E	1110	0	0000	1010
F	1111	7	0111	0000

and ANF(F1). Together they are called the LP of the respective ANF equation. Columns 9 through J of rows 2 through 5 give the value of the respective nonlinear coefficients more specifically, a_5 to a_{15} (according to Equation 2) of ANF(F4), ANF(F3), ANF(F2), and ANF(F1). Together they are called the NP of the respective ANF equation.

The fourth, third, second, and first IPVs for the given S-box have been noted in the "IPV" field under the columns labeled 4, 3, 2, and 1, respectively, from rows 8 through M of Table 10.8. The four OPBFs, F4, F3, F2, and F1, are noted in columns 4, 8, C, and G from rows 8 through M, respectively. The corresponding LP, NP, and satisfaction values (LP = BF) are noted in columns 5 through 7, 9 through B, C through F, and H through J from rows 8 through M, respectively, of Table 10.8.

Table 10.8 Linear Approximation Analysis

R|C	1	2	3	4	5	6	7	8	9	A	B	C	D	E	F	G	H	I	J
1		Coefficient		C			LP							NP					
2		ANF(F4)		1	1	0	1	1	0	0	0	1	0	1	1	0	0	1	0
3		ANF(F3)		1	0	0	1	1	1	1	0	0	1	0	0	1	0	0	0
4		ANF(F2)		1	1	1	0	0	1	1	1	1	1	1	0	0	1	1	0
5		ANF(F1)		0	0	1	0	1	0	1	1	0	0	0	0	1	0	0	0
6	I	IPV	S	F	L	N	S	F	L	N	S	F	L	N	S	F	L	N	S
7	D	4321	B	4	P	P	F	3	P	P	F	2	P	P	F	1	P	P	F
8	0	0000	E	1	0	0	1	1	0	0	1	1	1	0	0	0	1	0	1
9	1	0001	4	0	0	0	0	1	0	0	1	0	1	0	1	0	0	0	0
A	2	0010	D	1	1	0	0	1	1	0	0	0	1	0	1	1	0	0	1
B	3	0011	1	0	1	1	1	0	1	0	0	1	1	1	1	1	1	0	0
C	4	0100	2	0	1	0	1	0	1	0	1	1	0	0	1	0	1	0	1
D	5	0101	F	1	0	0	1	1	0	1	1	1	0	1	1	1	1	0	0
E	6	0110	B	1	1	1	1	0	1	0	1	1	0	1	1	1	1	0	0
F	7	0111	8	1	0	1	1	0	1	1	1	0	0	0	0	0	1	0	1

(Continued)

Table 10.8 (Continued)   Linear Approximation Analysis

R|C	1	2	3	4	5	6	7	8	9	A	B	C	D	E	F	G	H	I	J
G	8	1000	3	0	0	0	0	0	1	0	1	1	0	0	1	1	0	0	1
H	9	1001	A	1	1	0	0	0	0	0	0	1	0	1	1	0	0	1	1

R|C	1	2	3	4	5	6	7	8	9	A	B	C	D	E	F	G	H	I	J
I/D	IPV 4321	S/B		F 4	L/P	N/P	S/F	F 3	L/P	N/P	S/F	F 2	L/P	N/P	S/F	F 1	L/P	N/P	S/F
I	A	1010	6	0	0	0	0	1	1	1	1	1	0	1	1	0	0	1	1
J	B	1011	C	1	1	1	1	1	0	1	1	0	0	0	0	0	0	0	0
K	C	1100	5	0	0	0	0	1	1	1	1	0	1	1	1	1	1	0	0
L	D	1101	9	1	1	0	0	0	0	1	1	0	1	1	1	1	1	0	0
M	E	1110	0	0	0	0	0	0	1	0	1	0	1	1	1	0	1	1	1
N	F	1111	7	0	1	0	1	1	0	0	1	1	1	0	0	1	1	1	1

10.5.4 Result

No. of Linear Approximations with BF1	No. of Linear Approximations with BF2	No. of Linear Approximations with BF3	No. of Linear Approximations with BF4
7	4	2	8

Total number of existing linear approximations: 21.

10.5.5 Pseudocode with Time Complexity Analysis of the LAA Algorithm

The nonlinear part for the given analysis has been termed NP. The ANF coefficients are illustrated through the array anf [25]. IPVs are termed x_1, x_2, x_3, and x_4 for IPV1, IPV2, IPV3, and IPV4, respectively. The pseudocode of the algorithm of the above analysis is given below.

```
Start.
Step 1. NP = (anf[5].&x1&x2)^(anf[6]&x1 &x3)+(anf[7]&x1 &x4)+
(anf[8] &x2 &x3)+(anf[9]&x2 &x4)+(anf[10]&x3 &x4)(anf[11]&x1
&x2 &x3)+(anf[12]&x1 &x2 &x4)+(anf[13]&x1 &x3 &x4) +(anf[14]
&x2 &x3 &x4)+(anf[15]&x1 &x2 &x3 &x4))
Step 2. LP= anf[0] ^(anf[1].&x1)^ (anf[2].&x2)^ (anf[3].&x3)^
(anf[4].&x4).
Step 3. if(NP==0&& BF(x1x2x3x4) == LP) then Linear equation.
        else Nonlinear equation.
Stop.
```

Time complexity. Since the analysis contains no loops, the time complexity of the algorithm is O(n).

10.5.6 Comparison of Execution Time Complexity of Linear Cryptanalysis of 4-Bit Crypto S-Boxes and LAA of 4-Bit S-Boxes

The comparison of the time complexity of two algorithms is given in Table 10.9.

It can be concluded from the comparison that the execution time is reduced in LAA compared with the linear cryptanalysis of 4-bit crypto S-boxes. So, it can be concluded from the above review work that the execution time of the 4-bit linear approximation algorithm is much less than that of the 4-bit linear cryptanalysis algorithm, so the 4-bit linear approximation algorithm has been proved to be much better.

Table 10.9 Time Complexity Comparison of Two Algorithms

View	4-Bit Linear Cryptanalysis	4-Bit Linear Approximation
Time complexity	$O(n^2)$	$O(n)$

10.6 Analysis and Results

In this section, the analysis of results is described, as well as the security criterion for 4-bit crypto S-boxes.

10.6.1 Analysis of Results

The value of ^nC_r is maximum when the value of r is ½ of the value of n (when n is even). Here the maximum number of linear approximations is 64. So if the total satisfaction of a linear equation is 32 out of 64, then the number of possible sets of 32 linear equations is largest. This means that if the total satisfaction is 32 out of 64, then the number of possible sets of 32 possible linear equations is $^{64}C_{32}$. That is the maximum number of possible sets of linear equations. If the value of the total number of linear approximations with BF1 is close to 32, then it is more cryptanalysis immune since the number of possible sets of linear equations is too large to calculate. As the value nears 0 or 64, it reduces the sets of possible linear equations to search for, which reduces the effort to search for the linear equations present in a particular 4-bit S-box. In this example, the total satisfaction is 18 out of 64. This means that the given 4-bit S-box is not a good 4-bit S-box or a good cryptanalytically immune S-box.

10.6.2 Security Criterion for 4-Bit Crypto S-Boxes

If the values of the total number of existing linear equations for a 4-bit S-box are 24 to 32, then the lowest numbers of sets of linear equations are 250649105469666120. This is a very large number to investigate. So, the 4-bit S-box is declared a good 4-bit S-box or a 4-bit S-box with good security. If the value is between 16 and 23, then the lowest numbers of the sets of linear equations are 488526937079580. This not a small number to investigate in today's computing scenario, so the S-box is declared a medium S-box or an S-box with medium security. The 4-bit S-boxes having existing linear equations less than 16 are declared poor 4-bit S-boxes or vulnerable to a cryptanalytic attack.

10.7 Conclusion

In crowd computing, data transfer through the Internet is a vital issue. It is really hard to stop adversaries from interruption and destruction. Cryptographers made it possible to secure data transfer through the Internet. From the early days of commercial computer cryptography, cryptanalysis algorithms have been used to ensure the security of S-boxes as well as total encryption algorithms. The procedure of crowd computing has been of keen interest recently, but the security of transferred data through the Internet is the most vital issue in today's intelligent world. The encryption algorithms and S-boxes in these algorithms need to be secure from cyber-attacks although the algorithm and S-boxes are available to users. So, the security of nonlinear substitution or substitution boxes has been a major issue. From this analysis, we conclude that the algorithm is very lucid and efficient and secure for analyzing 4-bit S-boxes. The algorithm can easily be expanded to 8-bit, 16-bit, or 32-bit S-boxes. In crowd computing, the transfer of sourced data is a very important and intelligent work. The delivery of crowd-sourced data in a safe and secure manner is one of the most important roles of cryptology in crowd computing. The algorithm ensures the security of S-boxes and ultimately a secure algorithm for data encryption and decryption. The algorithm has also been proved to be less time-consuming and efficient for 4-bit S-boxes. It can also be concluded that the algorithm has also been proved to be able to provide security to 8-bit or 32-bit encryption standards.

References

1. E.J. Brown and W.A. Yarberry Jr. (2009). *The Effective CIO*. Boca Raton, FL: Taylor & Francis.
2. C. Shirky. (2011). *Cognitive Surplus: Creativity and Generosity in a Connected Age*. London: Penguin.
3. J. Surowiecki. (2005). *The Wisdom of* Crowds. New York: Random House.
4. B. Popper. (17 April 2012). Crowd computing taps artificial intelligence to revolutionize the power of our collective brains. *Venture Beat* (retrieved June 8, 2012). http://www.crowdcomputing.com/uses-and-examples-of-crowd-computing.
5. C. Adams and S. Tavares. (1990). The structured design of cryptographically good S-boxes. *Journal of Cryptology* 344(3), 27–41.
6. Data Encryption Standard. (1977). Federal Information Processing Standards Publication (FIPS PUB) 46. Washington, DC National Bureau of Standards.
7. M. Matsui. (1994). Linear cryptanalysis method for DES cipher. In *EUROCRYPT* 1994, no. 765, pp. 386–397.
8. M. Matsui. (1994). The first experimental cryptanalysis of Data Encryption Standard. In *Advances in Cryptology—CRYPTO '94*, 1994, pp. 1–11.
9. H.M. Heys. (2002). A tutorial on linear and differential cryptanlysis. *Cryptologia* 26, 189–221.
10. H.M. Heys and S.E. Tavares. (1996). Substitution-permutation networks resistant to differential and linear cryptanalysis. *Journal of Cryptology* 9, 1–19.
11. C.Y.Y. Design. (2000). Analysis and applications of cryptographic techniques. Department of Mathematics, Royal Holloway University of London.
12. A. Menezes, P. van Oorschot, and S. Vanstone. (1996). *Handbook of Applied Cryptography*. Boca Raton, FL: CRC Press.

13. B. Schneier. (1996). *Applied Cryptography*. 2nd ed. Hoboken, NJ: John Wiley & Sons.

14. E.F. Schaefer. (1996). A simplified data encryption standard algorithm. *Cryptologia* 20(1), 77–84.

15. B. Schneier. (2000). A self-study course in block-cipher cryptanalysis. *Cryptologia* 24(1), 18–34.

16. B. Schneier et al. (1999). *The Twofish Encryption Algorithm*. Hoboken, NJ: John Wiley & Sons. [The whole book is dedicated to different aspects of Twofish algorithm].

17. F. Mirzan. (2000). Block ciphers and cryptanalysis. Department of Mathematics, Royal Holloway University of London, pp.1–27.

18. H.M. Heys. (2000). A tutorial on linear and differential cryptanalysis. Memorial University of Newfoundland, Canada, pp.1–33.

19. H. Schulzrinne. (2000). *Network Security: Secret Key Cryptograph*. New York: Columbia University, pp.1–16.

20. L.G. Pierson. (2000). Comparing cryptographic modes of operation using flow diagrams. Sandia National Laboratories, Albuquerque, NM; Livermore, CA.

21. K. Aoki et al. (2000). Camellia: A 128-bit block cipher suitable for multiple platforms. NTT Corporation and Mitsubishi Electric Corporation.

22. S. Singh. (2000). *The Science of Secrecy*. London: Fourth Estate Limited.

23. S. Landau. (2000). Standing the test of time: The Data Encryption Standard. Sun Microsystems.

24. P. Garrett. (2001). *Making, Breaking Codes*. Upper Saddle River, NJ: Prentice Hall.

25. J. Kilian and P. Rogaway. (2000). How to protect DES against exhaustive key search. NEC Research Institute, USA.

26. B. Schneier. (2001). Why cryptography is harder than it looks. Counterpane Internet Security, USA.

27. K.S. Ooi and B.C. Vito. (2002). Cryptanalysis of S-DES. University of Sheffield Center, Taylor College.

28. J. Chen, D. Xue, and X. Lai, (2008). An analysis of international data encryption algorithm(IDEA) security against differential cryptanalysis. *Wuhan Univ. J. Nat. Sci.* 13(6): 697–701.

29. B. Schneier, J. Kelsey, D. Whiting, D. Wagner, C. Hall, and N. Ferguson. (1998). On the Twofish key schedule. *Fifth Annual Workshop on Selected Areas in Cryptography*. Berlin: Springer-Verlag, pp. 27–42.

30. L. Buttayan and I. Vajda. (1995). Searching for best linear approximation on DES-Like cryptosystems. *Electronics Letters* 31(11), 873–874.

31. J. Daemen, R. Govaerts, and J. Vandewalle. (1995). Correlation matrices. In: Preneel B. (eds) *Fast Software Encryption. FSE 1994. Lecture Notes in Computer Science*, vol 1008. Berlin: Springer.

32. E. Biham. (1995). On Matsui's linear cryptnalysis. In: De Santis A. (eds) *Advances in Cryptology — EUROCRYPT'94. EUROCRYPT 1994. Lecture Notes in Computer Science*, vol 950. Berlin: Springer.

33. C. Harpes G. Kramer, and J. Massey. (1995). A generation of linear cryptanalysis and the applicability of Matsui's pilling-up lemma. In: Guillou, L.C. and Quisqater, J.-J. (eds) *Advances in Cryptology—Eurocrypt '95*, pp. 24–38. Berlin: Springer.

34. B. Kaliski and M. Robshaw. (1994). Linear cryptanalysis using multiple approximations. In Desmedt Y.G. (eds) *Advances in Cryptology — CRYPTO '94. CRYPTO 1994. Lecture Notes in Computer Science*, vol 839. Berlin: Springer.

35. P.A. Junod. (1998). Linear cryptanalysis of DES. Zurich: Eidgenssische Tenhcische Hochschule.
36. B. Collard, F.X. Standaert, and J.J. Quisquater. (2008). Experiments on the multiple linear cryptanalysis of reduced round serpent. In K. Nyberg (ed.), *Fast Software Encryption. FSE 2008.* Lecture Notes in Computer Science, vol. 5086. Berlin: Springer.
37. L. Zhao, T. Nishide, A. Adhikari, KH. Rhee, and K. Sakurai. (2012). Cryptanalysis of randomized arithmetic codes based on markov model. In: Wu C.K., Yung M., and Lin D. (eds), *Information Security and Cryptology. Inscrypt 2011. Lecture Notes in Computer Science*, vol. 7537. Berlin: Springer.
38. M.A. Abdelraheem, J. Alizadeh, H. AlKhzaimi, M.R. Aref, N. Bagheri, and P. Gauravaram. (2015). Improved linear cryptanalysis of reduced-round SIMON-32 and SIMON-48. Cryptology E-Print Archive, Report-2015/988.
39. J. Feng, H. Chen, S. Gao, L. Fan, and D. Feng. (2016). Improved fault analysis on the block cipher SPECK by injecting faults in the same round. In *Proceedings of the 19th International Conference on Information Security and Cryptology*, Seoul, November 30–December 2.
40. X.L. Yu et al. (2015). Zero-correlation linear cryptanalysis of reduced-round SIMON. *Journal of Computer Science and Technology* 30, 1358–1369.
41. H. Feistel. (1971). Block cipher cryptographic system. U.S. Patent 3798359 (filed June 30, 1971).

Appendix

In this section, security analysis of 32 4-bit DES S-boxes has been carried out. The analysis is demonstrated in Table A.1, where LA indicates the number of existing linear relations in the S-box and LC represents the number of zeros in LAT. From analysis, except for a few marginal cases, results have been good in the new algorithm.

Table A.1   Security Analysis of Thirty-Two 4-Bit S-Boxes

S-Box	LA	LC
e4d12fb83a6c5907	21	143
0f74e2d1a6cb9538	29	143
41e8d62bfc973a50	23	138
fc8249175b3ea06d	25	154
f18e6b34972dc05a	24	132

(Continued)

Table A.1 (Continued)   Security Analysis of Thirty-Two 4-Bit S-Boxes

S-Box	LA	LC
3d47f28ec01a69b5	21	143
0e7ba4d158c6932f	31	143
d8a13f42b67c05e9	20	126
a09e63f51dc7b428	17	133
d709346a285ecbf1	22	133
d6498f30b12c5ae7	23	151
1ad069874fe3b52c	28	158
7de3069a1285bc4f	22	136
d8b56f03472c1ae9	22	136
a690cb7df13e5284	20	136
3f06a1d8945bc72e	22	136
2c417ab6853fd0e9	25	137
eb2c47d150fa3986	20	143
421bad78f9c5630e	30	130
b8c71e2d6f09a453	21	134
c1af92680d34e75b	30	141
af427c9561de0b38	29	127
9ef528c3704a1db6	24	127
432c95fabe17608d	24	130
4b2ef08d3c975a61	26	134
d0b7491ae35c2f86	27	145
14bdc37eaf680592	28	137
6bd814a7950fe23c	25	135
d2846fb1a93e50c7	23	144
1fd8a374c56b0e92	20	147
7b419ce206adf358	27	132
21e74a8dfc90356b	28	138

Index